Regeneration Code

The Creatures That Rebuild Themselves and the Mystery of Why We Can't

Team Trex

Published by NobleTrex Press

For permissions and other inquiries, write to:

P.O. Box 3132, Framingham, MA 01701, USA

Contents

Introduction

In the summer of 1740, a Swiss naturalist named Abraham Trembley peered through a magnifying lens at a sample of pond water and witnessed something that violated the fundamental rules of existence. He was looking at a small, green, trumpet-shaped creature attached to a weed-a freshwater polyp we now call *Hydra*. It looked like a plant, swaying in the current, but it moved with the volition of an animal. To settle the debate of its nature, Trembley did something drastic: he took a fine blade and sliced the creature in half.

If the world made sense, the creature should have died. A watch smashed with a hammer does not fix itself; a shattered vase does not reassemble. Entropy, the great unravelling, usually only moves in one direction.

But the *Hydra* did not die. Over the coming days, Trembley watched in disbelief as the bottom half grew a new head, and the top half grew a new foot. The single creature had become two, both perfect, both alive. He cut them again. Four creatures. He minced them into tiny pieces, and from the debris arose a battalion of new animals. It was a biological miracle that seemed to mock the fragility of the human condition.

Nearly three centuries later, we are still staring into that pond, haunted by the same question: *Why them and not us?*

We humans are the architects of the impossible. We have split the atom, walked on the moon, and cracked the genome. Yet, if you or I were to lose a finger in a workshop accident, the outcome is humiliatingly primitive. Our bodies will not unleash a cascade of architectural brilliance to rebuild the bone, the joint, the nail, and the fingerprint. Instead, we will panic. Our immune system will rush to the site not to rebuild, but to seal. We patch the breach with scar tissue-a biological welding job that prioritizes survival over function. We trade perfection for speed, and we live with the loss forever.

This is the great paradox of modern biology. We possess roughly the same gene families as the *Hydra*, the salamander, and the flatworm. The instructions for building a limb, a heart, or a lens are present in our DNA, used once during our embryonic development and then locked away, seemingly for good. We are walking around with a library of blueprints that our cells have forgotten how to read.

This book is the story of that forgotten library. It is an exploration of the "Regeneration Code"-the biological software that allows some creatures to rebuild themselves from scraps, and the investigation into the evolutionary padlock that keeps that software encrypted in humans.

For decades, the study of regeneration was relegated to the fringes of biology, viewed as a collection of "just-so" stories about worms and lizards-curiosities that had no bearing on human health. In the early 20th century, the father of modern genetics, T.H. Morgan, largely abandoned the study of regeneration in favor of the fruit fly, deciding that the problem was too messy, too complex, and perhaps insoluble.

But the tide has turned. Today, we are living through a renaissance of regenerative biology, driven by technologies that allow us to see the invisible. We can now sequence the genes of a single cell as it decides to become a neuron or a muscle fiber. We can watch, in real-time, as electricity sparkles across the wound of a frog, signaling the "all-clear" for regrowth. We are no longer just cataloging miracles; we are beginning to understand the mechanics behind them.

To understand why we break, we must first understand those who do not. In the pages that follow, you will meet the "Three Kings" of the regenerative world. You will meet the planarian flatworm, a cross-eyed marvel of the creek bed that is effectively immortal. A planarian can be cut into 279 pieces, and-as demonstrated by early pioneers and confirmed by modern labs-each piece will regenerate a new, complete worm. It is an organism that challenges our very concept of identity. If you are sliced in two and both halves regrow, which one is *you*?

You will meet the axolotl, the smiling Mexican salamander that serves as the champion of complex regeneration. While the flatworm is simple, the axolotl is a vertebrate, possessing bones, muscles, nerves, and blood vessels similar to our own. Yet, it can regenerate a lost limb perfectly, hundreds of times over, without ever forming a scar. It can even regenerate parts of its brain and heart.

And you will meet the zebrafish, a common aquarium pet that holds the secret to the Holy Grail of modern medicine: heart repair. While a human heart attack leaves behind a swath of dead, fibrotic tissue that eventually leads to heart failure, a zebrafish can have a chunk of its heart surgically removed and simply grow it back, beating and whole, within weeks.

The journey we are about to take is not just a tour of a biological freak show. It is a detective story. We are looking for the "Ghost in the Machine." As we analyze these creatures, we find that regeneration is not a magical event but a series of distinct, chemically driven decisions.

When a salamander loses a leg, its cells do something extraordinary. They stop acting like mature bone or skin cells and revert to a younger, more plastic state. They travel back in time. They form a structure called the *blastema*-a huddle of stem-like cells that acts as a construction site. This blastema is the crucible of regeneration. It consults a hidden map-a system of chemical coordinates and bioelectric signals-to determine exactly where it is and what is missing. It knows the difference between a wrist and an elbow. It knows when the arm is long enough, and it knows when to stop.

This brings us to the agonizing mystery of human biology. We do not form a blastema. We form a scab. Why?

For a long time, the prevailing theory was that we simply lacked the machinery-that evolution had discarded the regeneration genes as we became more complex. But recent evidence, particularly from comparative genomics, suggests something more unsettling. It's not that we *can't* regenerate; it's that we are actively prevented from doing so.

We will explore the "Faustian Bargain" of mammalian evolution. We will look at the possibility that our inability to regenerate is the price we pay for our warm blood, our complex immune systems, and, most crucially, our protection against cancer. Regeneration requires cells to divide rapidly and migrate wildly-behaviors that look terrifyingly similar to a malignant tumor. It appears that somewhere in the deep past, our ancestors evolved a biological kill-switch: better to scar quickly and survive to breed than to risk a runaway cellular process that could kill the organism. We traded the ability to regrow a limb for the ability to suppress tumors and survive in a bacteria-filled world.

But is that trade-off absolute? This is where the science gets truly exciting.

We will delve into the "Alchemist's Cell," exploring how scientists are learning to pick the lock on our own potential. We will discuss the role of the immune system, once thought to be the enemy of regeneration, now revealed as a critical foreman in the construction process. We will look at the work of researchers like those in Michael Levin's lab, who have shown that manipulating the bioelectric voltage of cells can trigger regeneration in species that shouldn't be able to do it. We will see how gene editing and the study of "zombie" cells-senescent cells that accumulate as we age-are reshaping our understanding of why healing slows down as we get older.

This book is also about the maps made of molecules. How does a cluster of cells know to build a thumb and not a pinky? The answer involves the beautiful mathematics of Alan Turing and the chemical gradients that provide a GPS for the body. We will see how these maps fade in humans, leading to the confusion of fibrosis, but how they might be redrawn.

We must also confront the reality of the "Wall of Scars." Fibrosis-the overproduction of connective tissue-is the end point of almost every chronic disease, from cirrhosis of the liver to kidney failure. It is the body's way of cementing over a problem it cannot solve. If we can crack the regeneration code, we don't just promise regrown limbs for amputees; we offer a potential solution to the organ failure that claims millions of lives annually.

However, I must offer a note of caution. This is not a book that promises you will regrow a leg next Tuesday by eating a specific superfood or thinking positive thoughts. The field of regenerative medicine has been plagued by hype, snake oil, and premature promises for centuries. We will look realistically at the "Promethean Horizon"-the gap between what is biologically possible in a dish and what is clinically possible in a patient. We will examine the ethical dilemmas of extending human life and health span. If we unlock the code to rebuild our bodies, do we also unlock the door to a radically different human experience?

The story of regeneration is the story of life's resilience. It is the story of how nature fights back against the second law of thermodynamics. Every time a lizard drops its tail and grows a new one, it is an act of defiance against the inevitability of decay.

As you read this, your own body is engaging in a quiet, low-level version of this war. The lining of your gut is replacing itself. Your skin is shedding and renewing. Your liver is maintaining its mass. You *are* a regenerating creature, just a limited one. The question is no longer whether we can expand those limits, but how far.

We are standing on the threshold of a new era in medicine, one where we stop simply patching the machine and start remembering how to rebuild it. The code is there, written in the nucleus of every cell in your body, waiting for the right signal to wake up.

Let us open the book and learn how to read it.

Chapter 1

Miracles in the Pond

In 1740, a Swiss naturalist cut a small polyp in half, expecting death, but witnessed the birth of two new lives, shattering the scientific dogma of the era. This chapter traces the shockwave of that discovery through to the modern laboratories where three specific creatures-the flatworm, the salamander, and the zebrafish-are finally revealing the biological code that allows them to rebuild body parts we can only mourn.

1.1 The Hydra's Defiance

A small greenish thread, no longer than a grain of rice, clings to a waterweed stem in a jar of pond water. If you lean close enough, you can just make out a crown of tentacles opening and closing like a slow, underwater hand. To an eighteenth-century naturalist, this creature was an invitation and an insult at the same time: an invitation because it was easy to keep alive on a windowsill; an insult because it refused to behave like a proper animal.

Abraham Trembley met this insult in the 1740s. He was not, by training, the kind of person who was supposed to rewrite biology. Trembley was a tutor in the Dutch Republic, employed in a well-to-do household, teaching children and tending a private curiosity for natural history. His laboratory was a domestic space: jars, pond water, bits of aquatic plants, and patience. Yet this was exactly the sort of setting in which a creature like the hydra could reveal its secret. The hydra did not demand an expensive instrument so much as an attentive eye and the willingness to do something that felt, at the time, almost indecently bold: to cut an animal apart and see what happened.

The freshwater hydra had already earned a reputation for confusion. It attached to plants like a polyp; it stayed put unless disturbed; it seemed to bud new individuals from its side, as if it were a plant making offshoots. But it also hunted. Tiny crustaceans and insect larvae that brushed those tentacles were seized and pulled to a mouth. That mouth, improbably, sat on the same end of the body as the tentacles, so feeding looked less like eating and more like the whole creature turning itself into a swallowing knot. The hydra was neither fish nor worm nor obvious insect. Even its name, borrowed from the many-headed monster of myth, hints at a sense that this was a creature built to embarrass tidy categories.

Trembley began with observation. He watched them feed; he watched them bud; he watched them contract into small knobs and then extend again into slim cylinders. The crucial step came when he decided not merely to watch but to intervene. If an animal was a kind of machine—as many influential thinkers were willing to suggest, at least metaphorically—then cutting it should behave like cutting a clock. You might remove a gear and the clock would stop; you might slice through the case and ruin the whole mechanism. Even if you believed animals were *more* than machines, you were still likely to assume that an animal's individuality was physically indivisible: one body, one life.

Trembley's knife violated that assumption with a calm that must have felt almost eerie. He cut a hydra in two. The two halves did not die. They did not even remain halves. Each portion, given time, reorganized itself into a complete animal. A headless piece formed a new head with tentacles; a tailless piece formed a new basal disc and anchored itself again. The point was not merely that the hydra could heal. Healing was already familiar: a wound closes, tissue knits, a scar forms. This was different. It was as if the creature contained an instruction not only for mending, but for *completeness*.

One of the most vivid episodes from Trembley's work involved turning the hydra into something like a biological puzzle. He did not stop at a single cut. He sliced them into multiple pieces; he cut off tentacles; he bisected them lengthwise; he tried to force them into unnatural arrangements. In some experiments he split the body column so that one hydra seemed to grow two heads. If the word *monster* had been hovering in the background as a mythic allusion, here was a monster made on a tabletop: not grotesque, not gigantic, but logically unsettling. If a creature could be encouraged to become two-headed, then the head was not a singular "seat" of life in the way common sense wanted it to be. And if each fragment could become a whole, then wholeness was not a privilege of the original body. It was a capacity distributed throughout it.

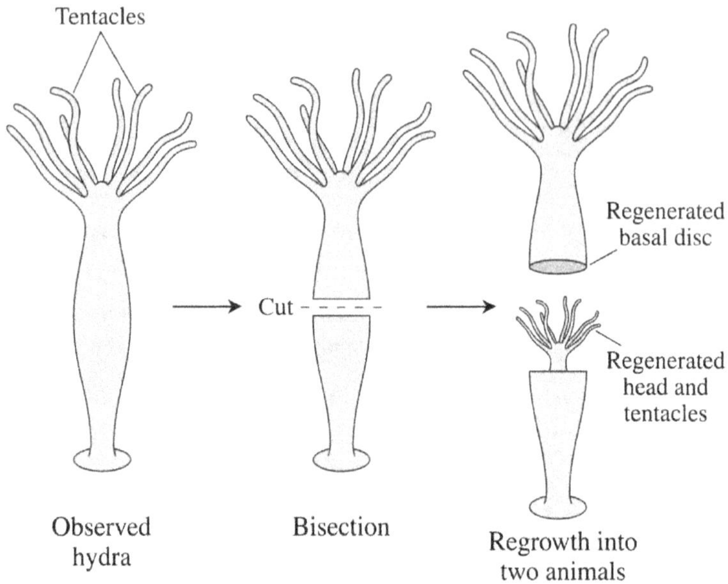

Trembley (1740s): cutting a hydra produced fragments that reorganized into complete animals.

Word travelled quickly in the Republic of Letters, that loose, international network of correspondence that linked scholars and gentlemen-naturalists across Europe. Trembley communicated his results to René Antoine Ferchault de Réaumur, a major figure in French science. Réaumur helped circulate the findings through the French Academy and beyond, turning a tutor's jar-bound observations into a public problem. The hydra became a test case not only for zoology but for philosophy. If anyone wanted a crisp definition of an animal, the hydra blurred it. If anyone wanted a neat boundary between plant and animal, the hydra trespassed across it.

The shock, for many readers, lay less in the creature's strangeness than in the implications of its resilience. To accept Trembley's hydra was to accept that living form could be *plastic*—capable of being reshaped and then re-established. "Plastic" here has nothing to do with polymers; it means malleable, able to take on new form. The hydra was not merely repairing a wound; it was reasserting an overall body plan after an upheaval that would have made most organisms simply perish.

It is hard, from a modern vantage point, to fully recover the sense of scandal. Today, schoolchildren learn that some animals regenerate. Yet in the mid-eighteenth century, the idea collided with several deep expectations.

One expectation was mechanical. The era was fascinated by automata—intricate machines that could mimic life well enough to delight and unsettle. If bodies were machines, then their parts should have fixed roles. A severed limb does not become a new person. A cut clock does not produce two clocks. Trembley's hydra, by becoming two hydras, seemed to mock the metaphor.

Another expectation was hierarchical. Many organisms appear organized around privileged regions: a head that senses and directs, a heart that pumps, a brain that coordinates. Even when people did not know the details of anatomy, they felt that some part must be *in charge*. The hydra made governance look strangely distributed. A fragment without a head was not "dead"; it was unfinished. And it knew, somehow, how to finish itself.

A third expectation was theological and moral. Cutting animals into pieces to test what life can endure raised questions about cruelty, about the soul, about what it means for something to be an individual. The hydra's indifference to injury complicated sentimental boundaries: if an organism can be divided and still live, what exactly is being harmed? This is not an argument for carelessness—the history of biology is full of harm. It is a reminder that regeneration experiments were, even at their birth, entangled with ethical discomfort as well as wonder.

Trembley's genius was to treat this discomfort not as a reason to stop, but as a reason to be careful. He repeated his manipulations, varied them, and described them with enough detail that others could follow. This is easy to say and difficult to do. Reproducibility, the ability for a result to survive beyond the original observer, is one of the quiet revolutions of experimental science. Trembley's hydra forced naturalists to stop thinking of living forms as static specimens and to start thinking of them as dynamic systems that could be perturbed and then watched as they responded.

The hydra's response also planted a conceptual seed that still shapes regeneration research: the difference between *rebuilding* and *repatterning*. Many animals regenerate by growing lots of new tissue. Think of a salamander regrowing a limb: a mound of cells forms at the stump, multiplies, and then differentiates—specializes into muscle, bone, skin—while the structure lengthens and shapes itself. Biologists later gave this growth-heavy strategy a name, *epimorphosis*, and the cell mass a name too: a *blastema*. The blastema is not a magic substance; it is a temporary, highly active zone of cells near the wound that behaves in some ways like a developmental bud.

Hydra regeneration looks different. Often it does not require dramatic growth. A fragment can reorganize its existing tissues, reshuffling proportions, re-establishing where the head should be and where the base should be, then growing only as needed to refine the final form. This strategy came to be called *morphallaxis*: the body pattern changes primarily through rearrangement and re-scaling rather than through building a large mass of new material. These terms are not labels Trembley used, but they capture what his hydra made vivid: "regeneration" is not one trick but a family of solutions.

The family resemblance hides a deeper mystery. How does a piece of animal know which end should become a head? If you cut a hydra in half, the fragment that used to be the middle can become an end. Somewhere in its tissues there must be a way to decide polarity: top versus bottom. Modern biology describes this kind of information as *patterning* or *positional information.* The phrase sounds abstract, but the intuition is everyday. When a city rebuilds after an earthquake, it does not merely replace bricks; it restores streets in the right places. Regeneration is not just replacing cells. It is restoring an arrangement.

Hydra make that point almost cruelly clear because they are so simple to describe and so hard to explain. Their body is essentially a tube with a mouth at one end and a sticky disc at the other, built from two main layers of cells with a jelly-like matrix between them. No bones, no heart, no brain in the vertebrate sense. Yet the rules that restore the body plan are precise enough that a piece becomes a coherent animal rather than a tumorous lump of tissue.

The temptation, when faced with such a creature, is to treat it as an exception—a toy organism, a biological party trick. But the hydra's defiance is more instructive than its simplicity suggests, because it exposes a general problem: animals are not merely heaps of specialized parts. They are self-maintaining patterns. Injury is not just damage; it is information the body interprets. The hydra is compelling because the interpretation is so ambitious.

The eighteenth-century debates sparked by Trembley's work were sometimes framed as arguments about what kind of creature the hydra really was. Plant or animal? Individual or colony? The framing mattered, because it determined whether the observations were revolutionary or merely odd. If a hydra was more like a plant, then budding and regrowth might seem less surprising. Plants can be propagated from cuttings; gardeners had long exploited that fact.

But the hydra's predatory behaviour and its muscular contractions resisted being pushed into the plant kingdom. The real lesson was that the categories themselves were too crude. There are animals whose bodies are organized in ways that permit what seems, by mammalian standards, almost casual self-duplication.

There is a second, subtler lesson in how Trembley's news spread. Trembley was not shouting into a void; he was writing to influential correspondents, and those correspondents were vouching for him. The hydra did not become a scientific problem solely because it was spectacular. It became a scientific problem because it arrived with a social apparatus: letters, endorsements, re-descriptions, replications. The modern reader may imagine science as a set of impersonal facts, but facts, especially disorienting ones, require communities willing to be disoriented together.

Hydra also created a productive embarrassment for the idea that an organism's form is fixed once it is adult. Adult bodies were often imagined as settled: development happens in the embryo, then the creature simply grows larger and eventually declines. Hydra insisted that development-like processes could be reawakened later, after injury, and not merely as crude repair but as organized rebuilding. If one wants a single phrase that captures the long shadow of Trembley's jar, it might be this: adulthood is not the end of biological creativity.

Modern molecular biology has given a sharper language to what Trembley could only describe with scalpel and patience. In hydra, signals that guide embryonic development in other animals also guide regeneration. One well-studied example is the Wnt signaling pathway, a set of molecular messages cells use to coordinate patterning, especially along head-to-tail axes. Wnt is ancient; versions of it exist across the animal kingdom. In hydra, injury can trigger Wnt-related activity that helps decide where a head organizer will form—

a region that behaves like a command post for building a head. The word "organizer" comes from developmental biology and refers to tissue that instructs neighbouring cells how to pattern themselves. Hydra, with their ability to regenerate heads again and again, make organizers feel less like a fleeting embryonic event and more like a reusable tool.

This modern language does not demote Trembley's achievement; it clarifies what was so conceptually explosive about it. The hydra did not regenerate because it possessed some special substance unknown to other animals. It regenerated by deploying common biological tools with uncommon freedom. That is precisely why it remains unsettling. If the tools are common, then the difference between regeneration and scarring cannot be reduced to whether a gene exists. The difference must lie in how, when, and where the tools are used, and how strongly other programs—wound closure, inflammation, tissue stiffening—override them.

Even within hydra, the story is not uniformly triumphant. Some hydra strains lose regenerative competence under certain conditions, and some show something like aging, with declining ability to maintain tissues. That nuance matters because it keeps the hydra from becoming a fairy tale. The capacity to regenerate is not a permanent badge; it is a biological state that can be supported or undermined. It depends on cellular housekeeping—processes like autophagy, the cell's way of recycling damaged components—and on the maintenance of stem-like cell populations that can replace lost tissues. In other words, the hydra is not exempt from biology's trade-offs; it simply sits at an extreme that makes those trade-offs legible.

Trembley could not have named stem cells or signaling pathways, but he understood something that still guides the best experiments: an organism can be interrogated by forcing it into a problem. Cut it, invert it, split it, and watch how it restores order. This is not vandalism; it is a way of asking, with unusual clarity, what the system considers "normal." The hydra's answer was not a shrug. It was a reassertion.

One can imagine, in a quiet room in the 1740s, the particular suspense of waiting. A cut is made. The jar is returned to the shelf. Hours pass; days pass. Under the lens, the fragment is neither obviously dying nor obviously becoming whole. Then, gradually, a swelling; the hint of a mouth; tentacles budding like the first shoots of a plant, except that they will sting and grasp and feed. The old creature is gone, not because it died, but because it became two.

That doubling carries a final provocation. When a hydra fragment becomes a complete hydra, it is not merely a repaired original. It is an individual with its own future, capable of budding offspring, capable of being cut again, capable of living as if the knife had been an inconvenience rather than an ending. The knife, meant to be a symbol of death, becomes instead a way of revealing a body's hidden logic: the ability to restore a pattern, to define itself again, to insist that "whole" is not a single moment in an animal's life but a state it can return to when the world makes a mess of it.

1.2 The Three Kings of Regrowth

Once you accept that an animal can rebuild itself, the obvious question becomes practical: where do you look to learn the rules? Not in the charismatic megafauna, and not in the creatures that happen to share our kitchens and cities. The best teachers are often small, unglamorous, and stubbornly reliable. Modern regeneration science has converged on three such teachers: the planarian flatworm, the salamander (especially the axolotl), and the zebrafish. They are not the only animals that regenerate, but they have become the working classics because each one excels at a different kind of biological comeback, and each one can be studied with tools that make mechanism—not just marvel—possible.

They are sometimes treated as mascots, but that misses their real value. Each animal forces researchers to confront a specific, hard problem. Planaria turn regeneration into a question about *pattern*: how do you rebuild an entire body, including a brain, and still end up with one head rather than two? Salamanders turn regeneration into a question about *integration*: how do you regrow a limb and weld it into a mature body so that nerves, muscles, and skeleton connect and function? Zebrafish turn regeneration into a question about *organ repair*: how do you replace living, working tissue like heart muscle without leaving a scar that compromises performance?

All three are also a rebuke to a quiet assumption that lingers in medicine: that adult tissues are basically stuck with what they've got. These animals are adults when they regenerate. They are not simply replaying embryonic development from scratch. They are doing something trickier: using some of the same genetic circuitry that built the body in the first place, but deploying it inside a body that is already built, already wired, already busy.

Planaria: the flatworm that treats amputation as a suggestion

If the hydra unsettled naturalists by refusing to stay singular, planaria unsettle biologists by refusing to stay incomplete. A planarian is a small freshwater flatworm, often a few millimetres to a couple of centimetres long, with a soft, leaflike body and a head that looks faintly hammer-shaped. If you look closely, you can see two dark eye spots. They are not eyes like ours—no lenses, no colour vision as we experience it—but they are enough for the worm to sense light and orient itself.

The famous claim about planaria is that you can cut one into pieces and each piece will regenerate a complete worm. That statement is true often enough to be useful, but vague enough to invite exaggeration. The real wonder is not that a fragment survives. It is that the fragment can rebuild *the right things in the right places.* A tail fragment does not simply grow random tissue until it resembles a worm-shaped blob. It builds a head at the correct end, complete with a brain (again, not like ours, but a genuine central nervous system), and it reconstructs a digestive system that branches through the body like a living river delta.

The biological engine behind this is a population of adult stem cells called *neoblasts*. A stem cell, in general, is a cell that can both make more of itself (self-renewal) and produce specialized descendants (differentiation). In humans, stem cells exist, but most are restricted: skin stem cells make skin; blood stem cells make blood. Planarians keep something closer to a general-purpose repair workforce in active service throughout life. Some neoblasts are *pluripotent*, meaning they can give rise to many different cell types. Pluripotent does not mean magical; it means wide-ranging potential under strict control.

That control is the hard part. Regeneration poses two problems that sound similar until you try to solve them. The first is *replacement*: you must generate the missing cell types. The second is *pattern*: you must know what is missing and where it belongs. If you amputate a head, the body needs neurons and sensory structures, yes, but it also needs to know to build *one head* at *one end* and to stop. Planaria succeed at both problems so reliably that they have become a kind of living logic puzzle.

For a long time, it was tempting to imagine neoblasts as a uniform soup of identical, all-powerful cells that simply poured into injuries and became whatever was needed. Modern single-cell technologies have made that picture less tidy and more interesting. Neoblasts are heterogeneous: there are subpopulations that already show biases toward particular fates, expressing transcription factors (genes that control other genes) associated with muscle, gut, epidermis, and other lineages. Yet fate bias does not necessarily mean fate lock-in. Work in the 2010s and 2020s, including Alejandro Sánchez Alvarado's lab and others, has helped establish a view in which planarian stem cells can begin to "lean" toward certain identities while still retaining the capacity to reset if the body's needs change. Regeneration, in this light, is not a single leap from potency to specialization. It is a controlled descent, with checkpoints.

A concrete example makes the abstraction feel real. Imagine a planarian cut straight across the middle. The front half has a head and brain but lacks a tail end; the back half has a tail but lacks a head. Both halves seal their wounds quickly. Then a structure called a *blastema* forms at the cut surface: a pale, cell-dense region where new tissue will be produced. The word blastema is used across regeneration biology, and it can mislead. It is not a uniform pile of identical cells. It is a site where cells are dividing, migrating, and receiving instructions. In planaria, those instructions include signals that mark "this end should be head" or "this end should be tail." The worm's success depends on keeping those signals consistent over time, so that the blastema does not start building a head and then change its mind halfway through.

Planaria became mainstream model organisms relatively late compared with fruit flies or mice, partly because their genetics and laboratory handling were less standardized for much of the twentieth century. But as methods improved, the worms' outrageous abilities became a disciplined experimental system. Their very plainness became a virtue: no fur, no bones, no complicated circulatory system. When a planarian solves the problems of replacement and pattern, it does so with a set of core biological tools that are easier to isolate and manipulate. That is why planaria are sometimes treated as a Rosetta Stone for regeneration. Not because they hold a single secret word, but because they make it possible to translate between the language of cells (division, differentiation) and the language of form (a head here, a tail there).

Salamanders: rebuilding a limb is the easy part; making it belong is harder

If planaria make regeneration look like mathematics, salamanders make it look like craft. A salamander limb is a complex structure: bone, cartilage, muscle, tendons, blood vessels, skin, and nerves, arranged in a precise three-dimensional architecture. In humans, losing a limb triggers healing that is vigorous and lifesaving but not restorative. Wound closure occurs, infection is fought, and a scar forms. In salamanders, the same basic urgencies exist, but the story continues. An amputated limb can regrow.

The axolotl, *Ambystoma mexicanum*, has become the emblem of this capacity. It is a salamander that retains juvenile features into adulthood (a trait called neoteny), including feathery external gills. Axolotls are not merely curiosities; they are a laboratory workhorse because they breed in captivity and tolerate experimental manipulation. They are also a conservation tragedy in the wild, but in the lab they have allowed scientists to turn limb regrowth from a natural history anecdote into a mechanistic discipline.

The classical picture of salamander limb regeneration involves a sequence. After amputation, the wound closes rapidly, forming a specialized wound epidermis. Under that epidermis, cells at the stump participate in forming a blastema. In salamanders, this blastema includes cells that arise in part through *dedifferentiation*—a process in which mature cells loosen their specialized identity and return to a more flexible state. Dedifferentiation does not mean they become embryonic all-purpose cells; rather, they may become progenitors that can rebuild related tissues. The details vary by cell type and by species, and ongoing research continues to refine how much true dedifferentiation occurs versus activation of resident stem cells.

However the blastema is assembled, it must do two things. It must generate the right tissues, and it must do so with correct geometry: shoulder to elbow to wrist to digits, with the proper orientation and scale. That second demand is easy to underestimate. Building tissue is one problem; building *anatomy* is another.

Even in salamanders, the anatomy is not always perfect. Recent work has emphasized that skeletal regeneration can be flawed: cartilage can form imperfectly, ossification (bone formation) can be irregular, and integration with pre-existing tissue can fail. That imperfection is not a disappointment. It is a clue. It suggests that regeneration is not a single switch labelled "on." It is a chain of steps, and any weak link can leave a limb that is present but not quite right.

The most surprising weak link, to many readers, is the immune system. In everyday language, immunity is what fights germs. In regeneration biology, immunity is also a kind of foreman. In a landmark study, James W. Godwin and colleagues (2013) showed that macrophages—immune cells that can engulf debris, secrete signals, and shape inflammation—are required early in axolotl limb regeneration. When macrophages were depleted at the wrong time, the wound still closed, but regeneration failed. Instead of a regrowing limb, the animals developed fibrosis: excessive connective tissue and extracellular matrix deposition, the biological substrate of scarring. Later, when macrophage populations recovered, re-amputation could restore regenerative ability, highlighting how strongly the early immune environment can set the trajectory.

This finding forces an uncomfortable comparison. Mammals are not uniquely bad at healing. We close wounds fast, and our immune systems are aggressive. That aggressiveness is protective. But in many contexts it comes with a tendency toward scarring. Salamanders, by contrast, seem able to mount an immune response that clears damage without pushing the tissue into a fibrotic dead end. The difference is not simply "less inflammation." It is timing, cell types, and the signals they exchange with fibroblasts (the connective-tissue-producing cells) and with the regenerating blastema. Regeneration is a negotiation, and macrophages are often in the room where it happens.

A historical anecdote gives this scientific point a human edge. Early experimenters were often amazed by salamanders' abilities and sometimes made them into parlour demonstrations. The spectacle of a limb regrowing was treated as proof that nature held hidden reserves of vitality. But the modern laboratory view is less romantic and more sobering. A regrowing limb is not a spontaneous miracle; it is the outcome of a coordinated program that can be disrupted by surprisingly subtle interventions. The axolotl does not "want" to regenerate in a sentimental sense. It regenerates because its cells, immune system, and tissues cooperate under a specific set of conditions that evolution has made possible.

Zebrafish: the heart that refuses to stay broken

If salamanders are the celebrities of limb regrowth, zebrafish are the quiet scandal of internal organ repair. A zebrafish is a small tropical fish with silver-blue stripes, common in aquaria and, more importantly, common in research facilities because it is easy to breed, produces transparent embryos, and has genetics that can be manipulated with precision.

The headline ability, often repeated because it feels so close to human hopes, is heart regeneration. In adult humans, a heart attack kills cardiac muscle cells (cardiomyocytes). The body responds with inflammation and then scar formation. The scar is not inert, but it does not beat. It alters the mechanical properties of the heart wall and can contribute to heart failure. Zebrafish can recover from substantial cardiac injury by replacing lost muscle and restoring function far more effectively than adult mammals.

The cellular logic differs from planaria and salamanders in a revealing way. Zebrafish do not rely on an all-purpose neoblast-like stem cell population to rebuild the heart. Instead, much of the repair comes from existing cardiomyocytes re-entering the cell cycle: mature heart muscle cells, normally reluctant to divide, become proliferative again. This is a theme in zebrafish regeneration more broadly: tissues often regenerate by persuading their own differentiated cells to temporarily behave like progenitors, rather than importing a separate stem cell army.

That persuasion is not cost-free. A dividing cardiomyocyte must loosen some of the structural specialization that makes it a good contractile cell. There is a tension between being excellent at your job and being able to reproduce. The zebrafish manages this tension with timing: cells de-specialize enough to divide, then re-specialize to restore function. The details are still being worked out, and even zebrafish can show transient scarring and imperfect maturation in some contexts. Again, the point is not to treat the animal as a flawless machine. It is to recognize a workable strategy that mammals largely suppress.

Zebrafish also regenerate other tissues: fins can regrow, and parts of the eye, including the retina, show remarkable repair capacity. The lens is a useful example because it highlights how regeneration is sometimes about restoring clarity and function, not just rebuilding mass. In humans, lens damage or clouding (cataract) is usually treated surgically, replacing the lens with an artificial one. Many animals regenerate lens tissue via coordinated responses of ocular cells. The zebrafish has become a model for understanding how eye tissues can be coaxed into regenerative programs, a subject with obvious medical stakes.

Why these three, and why not simply pick the "best" regenerator?

It is tempting to ask which animal is the champion. Planaria can rebuild a whole body; salamanders can rebuild a limb; zebrafish can rebuild a heart. But regeneration is not a single sport with a single trophy. The real value of these organisms is that they each make one aspect of regeneration experimentally tractable.

Planaria are unmatched for studying how a body keeps a map of itself. Salamanders are unmatched for studying how complex structures reform in three dimensions and reconnect with nerves and mature tissues. Zebrafish are unmatched for studying how a vertebrate organ can replace specialized working cells and regain function.

They also force a confrontation with human limits, because the comparison is not unfair. Humans share a great deal of genetic machinery with all three. We have Wnt signaling; we have stem cells; we have immune cells; we have developmental programs that once built a limb and a heart and a lens. The gulf is not explained by absence. It is explained by regulation: which programs get activated after injury, which are suppressed, and how forcefully scarring mechanisms dominate.

The simplest way to see that gulf is to compare outcomes organ by organ. The picture below is not meant to be exhaustive or perfectly universal across every experimental setup. It is meant to make a practical point: regenerative capacity is not evenly distributed across tissues, and humans sit at an extreme where scar-based repair often wins.

The matrix is blunt, and bluntness has a use: it clarifies the shape of the mystery. Humans are not uniformly non-regenerative. Skin renews constantly. The liver can restore mass after partial loss. Children can sometimes regrow fingertip tissue if the injury is just so. But when it comes to rebuilding complex, highly organized structures like a limb, or restoring non-scarring function after loss of heart muscle, adult human biology tends to choose stabilization over reconstruction.

The three model organisms make it harder to accept that choice as inevitable. Planaria show that adult bodies can keep a workable blueprint and redeploy it after catastrophic loss. Salamanders show that even complex anatomy can be rebuilt in a mature animal, and that the immune system can be a partner rather than an obstacle. Zebrafish show that specialized working cells can, under the right conditions, step backward just enough to divide and then step forward again into function.

Each animal also supplies a warning. Regeneration is not simply "more healing." It is a particular kind of healing that must be tightly governed. A system that allows cells to proliferate, migrate, and change identity carries risks that sound familiar in a different context: cancer, mispatterning, overgrowth, failure to integrate. Champion regenerators manage these risks. They do not abolish them. Their bodies walk a narrow ridge between flexibility and control.

There is a reason these creatures have become fixtures in laboratories rather than mere curiosities. They offer not only spectacle but leverage. You can silence a gene in a planarian and watch the worm rebuild incorrectly, revealing which molecular words matter for the body's grammar. You can remove a class of immune cells in an axolotl and watch the tissue choose fibrosis, revealing how scarring can become a default. You can track dividing cardiomyocytes in a zebrafish heart and see mature muscle re-enter the cycle, revealing that "post-mitotic" is not always a permanent sentence.

In the end, the most provocative thing about the three kings is not that they do what we cannot. It is that they do it with ingredients we largely share: cells that talk to each other, genes that can be switched on and off, and tissues that respond to damage with either imagination or caution. The difference is not a missing organ in the genome. It is a missing permission slip in the body's response to injury, and these animals keep handing scientists examples of what permission can look like when evolution decides to grant it.

1.3 The Missing Manual

A child scrapes a knee and, within days, the skin looks almost new. An adult scrapes a knee and, more often than not, a pale line stays behind like a quiet signature. The body has repaired the damage, but it has not restored the original. This is the ordinary human experience of injury: we heal, and we remember. In the animals that haunt the edges of regeneration research—flatworms that rebuild heads, salamanders that replace limbs, fish that patch hearts with living muscle—healing can look less like remembering and more like erasing.

It is tempting to explain the difference with a single, satisfying absence. Perhaps humans simply lack the necessary genes. Perhaps salamanders have a special molecular ingredient, a secret protein, a magical switch. Comparative genomics has made that story hard to sustain. Humans share most of the core genetic toolkit that builds bodies across the animal kingdom. We have Wnt, FGF, BMP, Hedgehog, Notch—the great, ancient families of signaling pathways that embryos use to lay out axes and organs. We have stem cell programs, immune programs, scar-forming programs. Our bodies can grow a limb once, in the womb, with exquisite patterning. The pieces are there.

The problem, then, is less like missing hardware and more like missing *instructions*. The "manual" is not a list of parts; it is a set of rules about when to deploy them, in what sequence, and with what constraints. An injury does not only break tissue. It forces the body to choose a response. Mammals often choose speed and closure. Many regenerating animals choose openness and rebuilding—not in the sense of leaving wounds exposed, but in the sense of keeping options alive long enough for a more ambitious reconstruction. That difference in priorities is the heart of the paradox.

One way to feel how deep the paradox goes is to consider fetal wound healing. In early mammalian development, wounds can heal with remarkably little scarring. The tissue seems to restore itself with a kind of quiet competence that adult skin has lost. By late gestation and into infancy, the same species shifts toward the adult mode: faster closure, more inflammation, more fibrosis, more scar. The genome has not changed. What has changed is the environment in which the genome is expressed: mechanical forces, immune tone, the behaviour of fibroblasts, the architecture of the extracellular matrix.

The extracellular matrix is the scaffolding and glue that holds cells in place—a meshwork of proteins like collagen. In everyday language, collagen is something we want in our skin. In wound healing, collagen can become a kind of emergency concrete. It stabilizes. It prevents rupture. It stops bleeding. But it also locks in a geometry that may be functionally adequate and aesthetically disappointing. In the heart, collagen-rich scar tissue can prevent the wall from tearing after a myocardial infarction, a heart attack. It also compromises the heart's ability to contract and conduct electrical signals cleanly. In the spinal cord, a glial scar can wall off damage and reduce spread of injury, but it can also obstruct axon regrowth. Scarring is not simply failure. It is a strategy with costs.

The cells most responsible for laying down this emergency concrete are *myofibroblasts*. A fibroblast is a connective-tissue cell that produces extracellular matrix. A myofibroblast is a more contractile, activated form: it pulls the wound edges together and deposits collagen. Myofibroblasts are excellent at rapid stabilization. They are also excellent at making scars persistent. Once myofibroblasts and their matrix become entrenched, they can create a physical and chemical environment that makes true regeneration harder. Cells trying to migrate or change identity must work through a stiff, fibrous terrain. Signals that promote scarring can drown out signals that promote patterning and growth.

There is a deceptively simple mechanical feature that distinguishes many mammalian wounds from the regenerative injuries of classic model organisms: *contraction*. Adult mammalian skin wounds often shrink dramatically as myofibroblasts pull the edges inward. This is efficient. It reduces the area that must be re-epithelialized (covered by new skin cells). In a small animal with loose skin, contraction can close a wound quickly. But contraction also changes the geometry of the tissue. It creates tension lines, distorts local architecture, and may help lock the body into "patch mode" rather than "rebuild mode."

Experiments that interfere with contraction can, in some contexts, reduce scarring and allow more regenerative outcomes. The details depend on the tissue and the animal model, and no one should mistake this for a simple clinical recipe. Still, the pattern is suggestive: the shift from fetal scarless healing to adult scarring correlates with the onset of stronger contraction and a different mechanical environment. The wound is not merely a biochemical event; it is a physical one. Cells feel tension. They respond to stiffness. They make decisions based on the forces they experience, not just on soluble signals floating in fluid.

This is one reason the "missing manual" metaphor is useful. A manual is not only written in words; it is written in the operating conditions a machine expects. Place the same device in a different environment—high pressure, low temperature, constant vibration—and it may behave differently even with the same internal components. A mammalian wound is a high-stress environment, flooded with inflammatory signals and dominated by urgent mechanical closure. A salamander limb stump, by contrast, quickly forms a specialized wound epidermis that seems to support blastema formation rather than short-circuit it. The salamander does not ignore infection risk; it orchestrates a different resolution.

The immune system sits at the centre of this orchestration. In everyday conversation, inflammation is treated as a villain. In reality, inflammation is a necessary first response. It clears microbes and debris. It recruits cells that will rebuild tissue. The question is not whether inflammation happens, but what form it takes and how it resolves.

Comparative studies have made this point uncomfortably clear. Some mammals regenerate better than laboratory mice. Spiny mice, for example, can regenerate skin and even parts of the ear pinna with reduced scarring compared with standard lab strains. Their wounds show different cytokine profiles—cytokines are immune signaling molecules—and different cellular dynamics. But attempts to reproduce a regenerative outcome in non-regenerating mice simply by reducing inflammation, for instance through COX-2 inhibition (a way of damping certain inflammatory pathways), have not magically turned scarring into regeneration. "Immune context" matters, but it is not a single lever. A wound's fate seems to be controlled by multiple interlocking layers: immune signals, mechanical forces, extracellular matrix composition, nerve supply, and developmental patterning cues.

The salamander macrophage story illustrates what "immune context" can mean in a more concrete way. In axolotls, macrophages arriving early after limb amputation are not merely cleanup crews. They appear to shape the extracellular matrix, influence fibroblast behaviour, and help create a permissive environment for blastema formation. When they are absent at the critical early window, the wound still closes, but it closes into fibrosis. Mammalian wounds, too, recruit macrophages, but macrophages come in different flavours and change over time. They can be pro-inflammatory, pro-fibrotic, pro-repair, pro-resolution. The balance and timing matter. A slight shift toward prolonged pro-fibrotic signaling can tip the wound toward a scar-stabilized endpoint.

Why would mammals evolve such a tendency? The blunt answer is that evolution does not optimize for perfect restoration. It optimizes for survival and reproduction in a given ecological context. Scarring is often "good enough," and sometimes better than the alternatives.

Consider infection. A large open wound is an invitation to bacteria and fungi. In a world without antibiotics, rapid closure can be lifesaving even if it leaves a scar. Consider bleeding. Mammals are warm-blooded and often active; they maintain high metabolic rates and, in many cases, high blood pressure relative to small ectothermic animals. Rapid stabilization of tissue can prevent fatal fluid loss. Consider predation. An animal that can quickly seal a wound and return to movement may survive to reproduce, while an animal that pauses to rebuild perfectly may become lunch.

Now consider cancer. Regeneration requires cells to proliferate, migrate, and sometimes loosen their identity. Those are also features of cancer. This does not mean regeneration causes cancer, nor that regenerating animals are cancer-ridden. Many of them manage proliferation with impressive discipline. But it does suggest a trade-off: a body that keeps adult tissues in a more plastic, development-like state might need stronger tumour-suppression strategies to prevent runaway growth. Mammals, with long lifespans and many dividing cells, face a significant cancer burden. Tight control over proliferation and rapid commitment to stable scar tissue could, in some contexts, reduce the risk of persistent, proliferative wound environments that might be exploited by malignant clones. The irony is that what looks like a limitation may also be a safety feature.

Another trade-off involves complexity and speed. Mammalian tissues are often highly specialized and densely integrated. Our skin is packed with appendages and nerves; our hearts beat billions of times; our brains rely on precise networks. A regeneration program must not merely create cells. It must recreate wiring, architecture, and function without destabilizing what remains. In some cases, the body may prefer a scar because it is predictable. A scar is a structural solution: patch the hole, restore barrier integrity, prevent catastrophic failure. Regeneration is a developmental solution: reopen a program that is powerful but harder to control in an adult context.

A historical anecdote helps illustrate how long these questions have been in the air. In the nineteenth century, as experimental biology matured, researchers were captivated by the ability of amphibians to regenerate. Limbs were amputated in careful series; tails were clipped; tissues were transplanted. The fascination was not only with the animals, but with the implicit insult to human medicine. Surgeons could suture and cauterize, but they could not persuade a limb to return. The gap between what the body *could* do in development and what it *would* do after injury felt like a locked door.

The locked door metaphor is apt, because many of the genes that drive regeneration in model organisms are not exotic. They are familiar developmental regulators, expressed in different patterns. The door is not made of missing bricks; it is made of decisions. After injury, mammalian tissues tend to activate programs that favour quick closure: clotting, inflammation, fibroblast activation, collagen deposition, contraction. Regenerating animals activate some of these too—they must stop bleeding and prevent infection—but they also create a permissive niche for patterning signals and proliferative progenitors. The "manual" is the choreography that lets those two needs coexist without one smothering the other.

This choreography is not only biochemical; it is also spatial. A scar is often a region of altered stiffness and altered cell composition. In mechanical terms, it changes the local "landscape." Cells behave differently on stiff substrates than on soft ones. They align, migrate, and differentiate in response to tension and geometry. A scar, once formed, can become a self-reinforcing state: stiff matrix promotes further myofibroblast activation, which deposits more matrix, which increases stiffness. Breaking that loop may be necessary for regeneration in mammals, and breaking it safely is a formidable challenge.

There is also an information problem hidden inside scarring. Patterning signals must be interpreted against a background. If the tissue around a wound becomes distorted by contraction and fibrosis, the gradients of signals—the subtle differences in concentration across space that cells can use to decide what they are—may become scrambled. A salamander blastema forms in an environment that seems designed to preserve positional information. A mammalian scar forms in an environment that prioritizes rapid sealing over faithful mapping.

Some of the most compelling evidence that mammals possess a "partial manual" comes from edge cases, where the body briefly behaves more like a regenerator. Mouse digit tips, under certain conditions, can regenerate. Human children sometimes show limited fingertip regrowth if the injury preserves specific structures. The details matter: the level of amputation, the presence of the nail bed, the extent of exposed bone. These edge cases suggest that mammalian tissues can, in narrow contexts, support a regenerative program. The question becomes why that program is so context-dependent, and why it shuts down so abruptly outside those boundaries.

One answer is that regeneration requires a permissive combination of signals that mammals rarely allow to persist. Another is that regeneration requires certain cell populations that are abundant or accessible in some contexts and scarce in others. Yet another is that regeneration requires nerves. In salamanders, nerve supply is classically required for limb regeneration; denervated limbs regenerate poorly or not at all. Nerves are not just wires; they are sources of growth factors and regulators of immune and vascular responses. Mammalian injuries can involve nerve damage and altered nerve signaling, potentially removing a key pro-regenerative cue.

None of these explanations are satisfying on their own, which is exactly the point. The missing manual is not a single missing page. It is a set of interdependent instructions. Change one element and the system may compensate by doubling down on scarring. Reduce inflammation and the body may still contract. Reduce contraction and fibroblasts may still deposit stiff matrix. Encourage proliferation and tumour-suppression mechanisms may apply the brakes. The wound response is a network of decisions, not a linear pipeline.

This network view also changes the emotional meaning of regeneration research. It is easy to treat regenerating animals as sources of envy and to treat human limitations as a kind of betrayal by nature. But the mammalian strategy is not simply stinginess. It is a particular compromise struck in the face of infection, bleeding, cancer risk, and the need for reliable function. The real question is whether that compromise can be renegotiated locally—in a wound, in a heart after infarction, in a spinal cord lesion—without rewriting the entire body's safety systems.

If that renegotiation is ever going to happen, it will not come from importing "new" genes as if humans were missing a part. It will come from learning how regenerating animals sequence their responses: how they clear damage without cementing a scar, how they permit cells to proliferate without losing control, how they preserve or recreate positional information, how they resolve inflammation into rebuilding rather than into fibrosis. The hydra's defiance showed that bodies can reassert wholeness. The planarian, salamander, and zebrafish show that wholeness can be reconstructed in different ways. The unsettling implication is that the barrier in mammals is not possibility, but governance: the body's preference for the quick, safe patch over the slower, riskier restoration.

A scar, in this light, is not merely a mark of damage. It is the fossil of a decision the body made under pressure: close fast, stabilize, and move on. The missing manual is the set of circumstances under which the body would dare to make a different choice.

Chapter 2

The Alchemist's Cell

When a salamander loses a limb, it does not merely patch the wound; it summons a localized cellular time machine capable of reversing biological age. By contrasting the immortal stem cell reserves of the flatworm with the de-differentiating tissues of the newt, we uncover the two distinct strategies nature uses to conjure new form from old matter.

2.1 The Construction Site

A wound looks like an ending. Blood, exposed tissue, the blunt fact of loss. In a mammal, the body treats that moment as an emergency to be contained: stop the leak, seal the breach, wall it off, move on. That response is fast and usually lifesaving. It is also, for regeneration, a kind of closing argument. A scar is not merely a patch; it is a decision.

In a salamander, the same moment is handled with a different logic. The wound is still an emergency, but it is also a work order. The animal does not simply *cover* the damage; it builds a temporary organ whose sole job is to manage rebuilding. Biologists call this organ the *blastema*. The word comes from the Greek *blastos*, meaning a bud or sprout, and it is a good name: the blastema is a living bud that grows where an arm or tail used to be.

Two things make the blastema easy to misunderstand. First, it is not a single cell type or a fixed structure like a liver. It is a shifting collaboration: epidermal cells at the surface, immune cells moving in and out, nerves delivering signals, and beneath it all a dense crowd of dividing progenitor cells (cells capable of producing other cell types) that will become bone, muscle, tendon, and skin. Second, it is born from a paradox. The animal must *avoid* the very processes we celebrate in human medicine—rapid clotting and collagen-rich scar formation—long enough to keep the stump flexible, permissive, and information-rich. Regeneration begins not with knitting the wound shut, but with keeping it in a carefully managed, unusually plastic state.

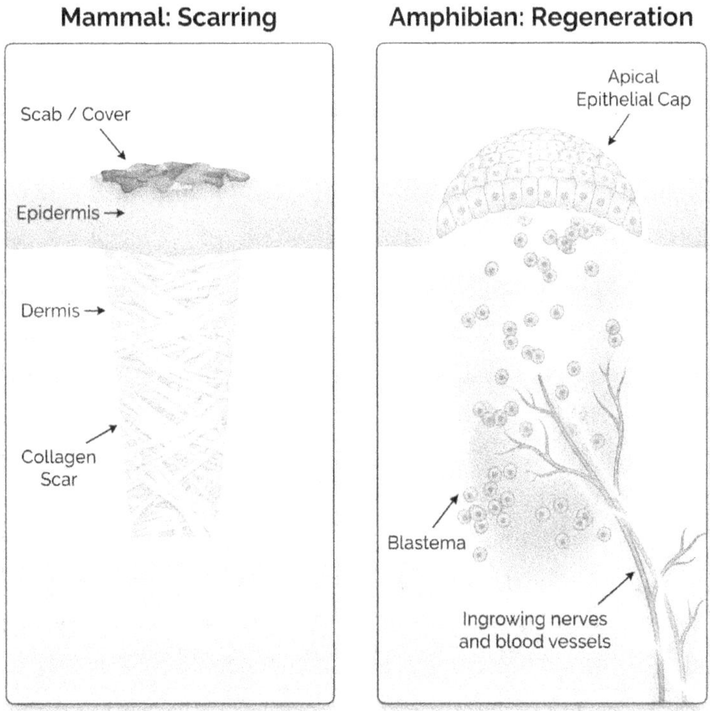

The contrast is easiest to see at the surface. In a typical mammalian wound, the epidermis (the outer skin layer) races to close over the gap, while fibroblasts in the dermis lay down collagen, a tough rope-like protein that gives skin its strength. Collagen is wonderful for durability. It is also, when deposited quickly and densely, a form of biological rebar: it stiffens the tissue, narrows the options, and locks the wound into a stable geometry. A scar is a strong compromise. It trades flexibility for security.

In salamanders, the surface does not behave as a simple lid. Within hours, epidermal cells migrate to cover the wound, but instead of settling into ordinary skin they become a specialized *wound epider-*

mis. As regeneration proceeds, this wound epidermis thickens into a structure called the *apical epithelial cap* (AEC). The AEC is a cap in the literal sense—a thickened pad of epidermis sitting over the stump—but it is also a cap in the managerial sense: it is an organizer, a foreman with a clipboard, issuing signals that tell the cells beneath it to proliferate, to stay youthful, and to keep their options open.

The body beneath the cap must be made receptive to rebuilding. That requires a second shift: the stump has to be remodeled into a site where cells can move, divide, and listen. This is where the extracellular matrix becomes central. The extracellular matrix is the meshwork of proteins and sugars outside cells that gives tissues their structure. In everyday life it acts like architecture: it holds cells in place and, through physical and chemical cues, helps keep them behaving like skin cells, muscle cells, cartilage cells. After an amputation in a regenerating animal, parts of this architecture are actively dismantled. Enzymes called matrix metalloproteinases (MMPs) increase early and chew through existing matrix components. New matrix components associated with regeneration appear, including fibronectin and tenascin, which make a scaffold that is less like a rigid wall and more like temporary plywood and scaffolding: firm enough to support activity, loose enough to rearrange.

That word *temporary* matters. A mammal's scar can persist for life. The blastema is meant to be spent.

At the center of the construction site is the crowd of cells that gives the blastema its name. Under the AEC, cells accumulate that look, under a microscope, like small, tightly packed, relatively undifferentiated cells. "Undifferentiated" does not mean empty or primitive; it means they have not yet committed to being one specialized adult cell type. They are proliferative. They are responsive. They are the raw workforce. But where do they come from?

The intuitive, almost fairy-tale answer is that mature tissues melt back into a soup of stem cells, and that soup becomes the new limb. Early accounts of salamander regeneration sometimes spoke as if the blastema were a uniform mass of magically reset cells. Later experiments complicated that story in a way that is more interesting and, strangely, more plausible. When researchers labeled specific tissues in axolotls (a salamander famous both in laboratories and in Mexican lakes) and traced their descendants through regeneration, many of those tissues contributed primarily to their own lineages. Muscle made muscle; cartilage contributed to cartilage; skin-derived connective tissue behaved like connective tissue. The blastema, rather than being a single egalitarian pool, behaved more like a mixed workforce where different trades show up at the same site. Carpenters do not become electricians, but both are essential to finish a house.

This is a subtle change of emphasis with big implications. It suggests that regeneration does not require complete cellular amnesia. Cells can keep a memory of their origin and still participate in rebuilding, provided the environment gives them permission to proliferate and the right instructions for where to go and when to stop. The blastema is less a vat of possibility than a tightly managed project in which many specialists are temporarily willing to behave like apprentices again.

What makes a stump willing? One answer is surprisingly social: the immune system.

The immune system has a reputation as the body's bouncer and executioner. In wounds, immune cells swarm in, kill microbes, clear debris, and orchestrate inflammation. In mammals, that inflammatory phase is coupled to scarring. In salamanders, the immune response still arrives, but the relationship between inflammation and outcome looks different. When researchers experimentally depleted macrophages (a key immune cell type that engulfs debris and secretes signaling molecules) during the early window after limb amputation, adult salamanders failed to regenerate properly and instead formed fibrotic, scar-like stumps. Even when macrophages later returned, regeneration did not simply resume as if nothing had happened. Timing mattered. The early immune environment was not just cleanup; it was part of the decision-making machinery that determines whether the tissue becomes a construction site or a permanent barricade.

That finding has the satisfying harshness of a good experiment: remove one worker, and the entire build fails. It also hints at why our own wounds tilt so predictably toward scarring. Human macrophages are not absent, of course. The difference seems to lie in the choreography: the sequence of signals, the balance between inflammatory and pro-resolution programs, and the way those signals interact with fibroblasts and the extracellular matrix. In a regenerating limb, immune cells appear to help create a permissive matrix and a pro-growth environment without letting collagen deposition harden too soon. The body is still defending itself; it is defending itself *in a way that keeps the future negotiable.*

Another answer is electrical, in the plain sense that cells carry voltages.

Every living cell maintains an electrical potential across its membrane: a tiny voltage created by pumping ions in and out. Collections of cells can create patterns of voltage differences, and those patterns can influence behavior: division, migration, gene expression. In a wound, the surface epithelium acts like an electrical sheet, and injury can create currents and fields that guide cells. Regenerating animals appear to use these bioelectric cues as part of their coordination toolkit. Gap junctions (channels that directly connect neighboring cells, allowing ions and small molecules to pass) help tissues behave like a synchronized group rather than isolated individuals. In the context of regeneration, the wound epidermis and AEC are not passive coverings but active, electrically and chemically communicative tissues.

This is not mystical electricity. It is closer to the way a city's power grid constrains what can be built where and when. You can pour concrete at midnight if the lights are on, if the permits are granted, if the trucks have a route. Bioelectric patterns are one of the permitting systems cells use to decide what actions are allowed.

The most dramatic permitting system, though, is neural.

A classic observation, known for well over a century in various forms, is that salamander limb regeneration is nerve-dependent. If the limb is denervated (its nerves severed), the stump may heal but the regenerate fails to grow properly. The nerve is not merely a cable that carries sensation; it becomes a supplier of growth-promoting factors and organizational signals. In modern terms, the nerve helps maintain the AEC and blastema in a proliferative state. Without nerve-derived signals, the construction site loses its foreman and its payroll.

One historical anecdote makes the point with the blunt clarity of early experimental biology. In the late nineteenth and early twentieth centuries, researchers working with newts and salamanders performed denervation experiments that were, by current standards, almost brutally simple: cut the nerves, amputate, observe. Over and over, the pattern held. The wound would close, but the limb would not return. To scientists of that era, this was not just a technical detail; it was a philosophical provocation. It suggested that regeneration was not an automatic replay of development. It was a negotiated process requiring ongoing information from the nervous system, a structure that in most textbooks was treated as a receiver of information, not a generator of anatomical instructions.

Later work put molecules to this old observation. Growth factors and signaling proteins induced by nerves, and sometimes produced in the wound epidermis in response to nerves, help sustain blastemal cell division. This also gives a human reader an unsettling thought: our own nerves are perfectly good at delivering signals. Yet our tissues interpret injury as a cue to scar. The problem is not a lack of communication hardware; it is the operating system.

The blastema, then, is not simply "cells that divide." It is a temporary organ that links epidermis, immune system, nerves, and matrix remodeling into a coherent program. The program has a rhythm.

First comes coverage: epidermal migration to seal the surface quickly, reducing infection risk while avoiding the heavy collagen barricade of scarring. Then comes specialization: the wound epidermis thickens, becomes the AEC, and begins to emit signals. In parallel, the stump undergoes controlled demolition: extracellular matrix is degraded and replaced, and differentiated tissues near the cut edge loosen their rigid identities enough to contribute progenitors. Immune cells arrive not only to clean but to tune the environment, preventing chronic inflammation and steering fibroblasts away from permanent fibrosis. Nerve signals act as a growth license, keeping the AEC functional and the blastema proliferative.

Only then does the visible outgrowth begin. Under the AEC, the blastema expands, and the stump develops a bulge. It is, at first, generic: a rounded mass of cells. What is astonishing is that this apparently simple bulge contains the information needed to rebuild complex anatomy with appropriate scale and proportion. A wrist does not become a shoulder; a hand does not sprout at the elbow if the cut was at the forearm. This implies that the construction site is not simply stocked with bricks; it is stocked with a plan.

The "plan" is not a little blueprint tucked inside a single cell. It is distributed. Some of it lives in gradients of gene expression along the limb's axes. Some of it is carried by connective tissue cells that retain positional identity. Some of it is enforced by interactions between cells as they sort themselves and respond to signals from the AEC. And some of it is supplied by the environment: mechanical forces, electrical patterns, nerve-derived factors. In other words, the blastema is not an object; it is a relationship.

That relationship can fail in instructive ways. If the early window is missed—if the immune response is derailed, if the wound epidermis does not specialize properly, if nerves are absent, if matrix remodeling locks the stump into stiffness—the body will default to the safer program. The wound closes, collagen accumulates, and the tissue becomes stable. Stability is the enemy of change.

Seen in this light, the most mysterious feature of the blastema is not that salamanders can grow limbs. It is that they can keep a wound open to possibility without letting it become chaos. They avoid infection and hemorrhage, yet resist premature sealing. They encourage proliferation, yet do not tip into uncontrolled growth. They dismantle tissue architecture, yet do not collapse into disorganized mass. A construction site is always dangerous. It requires fences, supervisors, schedules, and inspectors. Biology's inspectors are unforgiving: too much inflammation and you get fibrosis; too little and you get infection. Too much proliferation and you invite cancer; too little and the limb stalls. The blastema exists in that narrow band where risk is tolerated because the payoff is anatomical renewal.

At the end of the process, the most revealing thing happens: the blastema disappears. Not by dying dramatically, but by being used up. Cells that were proliferative commit to fates. The extracellular matrix shifts from provisional to permanent. The AEC thins back into ordinary epidermis. Nerve supply and blood vessels integrate into the new tissues, no longer as special instructors but as ordinary infrastructure. What remains is not a scar with a story; it is a limb that looks, moves, and feels like it belonged there all along.

2.2 The Immortal Reserve

A salamander has to *make* a regeneration organ. A planarian carries one everywhere.

Planarians are small freshwater flatworms, often no bigger than an eyelash, that glide over stones and leaves with a calm, slow confidence. They look simple, almost minimal: a soft ribbon of tissue with a head end that sometimes seems to squint back at you through two dark eyespots. Then you learn what they can do, and the simplicity begins to feel like misdirection. A planarian can be cut into pieces and, under the right conditions, each piece can rebuild the whole animal. This isn't limited to neat, generous halves. In many species, remarkably small fragments can regenerate a complete worm with a brain, gut, muscles, and skin. The feat is so robust that it has shaped the culture of biology: planarians became a kind of laboratory campfire story, a creature that could make even hardened experimentalists briefly sound like children describing a magic trick.

The trick has a name, and it is both less magical and more unsettling: *neoblasts*. Neoblasts are adult stem cells—cells in an adult body that can divide and produce specialized descendants. Many animals have adult stem cells. Humans rely on them constantly: blood stem cells in bone marrow, stem cells in the gut lining, stem cells in the skin. The planarian difference is not that it has stem cells, but that it has them in a way that feels almost extravagant. Neoblasts are distributed through much of the body, kept in a state of readiness, and capable of making *all* cell types. "Pluripotent" is the technical term for this breadth of potential: a pluripotent cell can give rise to many different tissues. In planarians, at least a subset of neoblasts appear genuinely pluripotent even in adulthood, not merely during embryonic life.

That last phrase—*in adulthood*—is the hinge on which the story turns. Most animals lock pluripotency away early. We allow it briefly, during development, because you cannot build a body without cells that have not yet chosen what they will become. Then, as the organism matures, pluripotency becomes a risk. A cell that can become anything may also become something you do not want: a tumor. Mammals, in particular, are cautious. We keep most adult stem cells on tighter leashes, more committed, more restricted in what they can make. Planarians take the opposite gamble. They keep an "immortal reserve" of powerful stem cells and build their biology around managing the danger.

Even the word "reserve" can mislead, because it suggests a stockpile hidden behind a fence, to be opened only in crisis. Planarian neoblasts are not sleeping. They are busy. A planarian is constantly replacing cells lost to wear, injury, and ordinary turnover. If you starve a planarian, it shrinks dramatically; if you feed it, it grows. Either way, it maintains proportion: the head remains a head, the body remains a body, the organs scale down or up. That scaling depends on an ongoing balance between cell division and cell loss, and neoblasts sit at the center of that balance. Regeneration is not a special mode that switches on after catastrophe; it is an acceleration and redirection of a program already running.

A useful way to picture this is to imagine two different kinds of repair culture. In one, your city has no spare workers on the payroll. When a bridge collapses, you assemble a crew, retrain people, redirect resources, and hope the economy can tolerate the disruption. In the other, the city maintains an always-employed workforce capable of building any kind of structure, and the main challenge is telling them what to build *today*. Planarians live in the second city. Their problem is not recruitment. Their problem is instruction.

A historical anecdote makes the strangeness concrete. In the late nineteenth century, the zoologist Thomas Hunt Morgan—long before he became famous for fruit-fly genetics—studied regeneration in planarians. Working with simple tools and a sharp blade, he cut worms in different ways and described the outcomes with a mix of precision and disbelief. This was a time when biology still argued, sometimes heatedly, about what "regeneration" even meant: was the new structure assembled from existing parts, or did it arise from something like a latent embryonic program? Morgan's planarian experiments did not solve everything, but they forced the question into the open. A fragment that can generate a whole animal is not merely repairing damage; it is reorganizing identity. It is answering the question, "What am I now?" with a body.

For a long time, neoblasts were defined by what they looked like: small cells with a large nucleus and scant cytoplasm, scattered through the planarian body but absent from the most anterior tip of the head and the pharynx region. That absence became a clue. When planarians were exposed to X-rays, they lost their regenerative ability and eventually died as their tissues failed to renew. Under the microscope, the neoblast-like cells vanished. Radiation kills rapidly dividing cells, and neoblasts divide. The correlation was suggestive: neoblasts seemed to be the dividing population that made regeneration possible. Yet correlation is a weak form of proof. Many things divide. Many things die under radiation. The real question was sharper: is there a single cell in the adult worm that can repopulate the entire organism?

In 2011, Peter Reddien's lab and colleagues provided a definitive answer using a beautifully direct experiment. They irradiated planarians to eliminate neoblasts, producing animals that could not regenerate and would eventually fail. Then they transplanted neoblasts from healthy donors. In some cases, a single transplanted cell was enough to rescue the host, restoring tissue turnover and the ability to regenerate. That single cell expanded, produced descendants, and rebuilt the stem cell compartment. This is one of those results that, once you grasp it, changes the meaning of the creature in your mind. A single adult cell, in the right context, can rebuild an entire animal's cellular economy.

The rescued worms were not merely patched. They regained the capacity for ordinary life. The transplanted cell did what a true stem cell must do: *self-renew* (make more of itself) while also producing specialized progeny. The term "cNeoblast" is often used for such clonogenic neoblasts—cells capable of forming a clone that sustains the animal. The experiment did not claim that every neoblast is equally potent. It argued something more precise and more powerful: within the neoblast population exists a subset with bona fide pluripotency.

Once that door was opened, the next wave of questions became molecular. If some neoblasts are pluripotent, how do you recognize them before transplantation? What genes do they express? How are they protected from genomic damage, given that they divide so often? Here modern techniques—especially single-cell sequencing—reshaped the field. Instead of averaging gene expression across whole tissues, researchers could profile individual cells and see the diversity within "the neoblasts." The population that once seemed uniform under a microscope broke into subtypes, including cells already biased toward particular fates and cells that looked more like a core, undifferentiated pool.

Marker-based isolation followed. Using cell-surface markers, including proteins such as TSPAN-1 in combination with high levels of stem-cell-associated transcripts like PIWI-1, researchers enriched for neoblast subsets with strong regenerative capacity. This kind of work matters not because it adds jargon, but because it changes what can be tested. If you can isolate a defined population, you can ask what signals maintain it, what stresses destroy it, and what genetic programs distinguish a truly pluripotent neoblast from a progenitor already leaning toward muscle or gut.

The PIWI family deserves special attention, because it connects regeneration to a deeper biological anxiety: the stability of the genome.

PIWI proteins are best known from animal germlines, where they help silence transposable elements—genetic sequences that can copy and insert themselves elsewhere, sometimes damaging genes or destabilizing chromosomes. The germline must protect DNA because it is passed to offspring. Planarians, in effect, treat their neoblasts with a similar seriousness. Genes such as *smedwi-2* (a planarian PIWI-family gene) are essential for neoblast function and regeneration. When PIWI pathways are disrupted, neoblasts fail, regeneration collapses, and the animal cannot maintain its tissues. The implication is both straightforward and profound: if you are going to keep pluripotent, highly proliferative cells in an adult body, you need a robust system to keep their genetic machinery clean.

This begins to answer an unspoken question that shadows every discussion of planarian regeneration: why don't planarians dissolve into cancer?

Cancer is not a single disease but a family of failures in growth control, often requiring multiple mutations and the ability to evade tissue constraints. A worm packed with dividing, powerful stem cells seems like it should be a tumor factory. Yet planarians are not known for rampant neoplasia in the way one might fear. Their biology suggests that proliferation and safety can coexist, but only with strong governance: genome protection mechanisms, differentiation programs that reliably pull cells out of the stem state, and tissue signals that keep growth coordinated with form. The neoblasts may be immortal in potential, but they are not anarchic.

Another ingredient of governance is time itself. Every division risks shortening telomeres, the repetitive DNA caps at chromosome ends that tend to erode as cells replicate. In many organisms, telomere shortening is part of the story of aging and limits on cell proliferation. Planarians complicate this story, and they complicate it in a way that fits their lifestyle. Some planarians reproduce sexually; others commonly reproduce asexually by fission, splitting and regenerating. In asexual lines, the animal's body effectively becomes its lineage. If telomeres shortened inexorably, long-term asexual propagation would be threatened. Studies have found that planarians can maintain telomere length through ongoing somatic proliferation in ways that depend on reproductive mode, supporting the idea that their stem cell systems are built for endurance. A planarian does not merely regenerate once; it can do so repeatedly, across events that would age and exhaust many other tissues.

The more one learns about neoblasts, the more planarian regeneration begins to look like a problem of logistics rather than raw ability. If the body already contains cells that can make everything, the hard part is instructing them to rebuild correctly: head where the head belongs, tail where the tail belongs, organs in the right proportions, symmetry maintained, injuries repaired without duplicating structures or leaving gaps.

Planarians solve that by combining the immortal reserve with a second kind of system: signaling networks that encode body pattern. The details can become technical quickly—gradients of Wnt signaling for head-tail identity, BMP signaling for back-belly patterning, and other pathways borrowed from embryonic development. What matters for an intuitive understanding is that neoblasts are not miniature worms-in-waiting. They are versatile labor, and the planarian body provides a constantly updated map.

That map is not written in words; it is written in molecular conditions. A neoblast near the head experiences a different cocktail of signals than a neoblast near the tail. After amputation, the edges of tissue change what they emit, and the signals reorganize. Neoblasts proliferate, but their daughters interpret the local environment and commit to the fates needed there. In this sense, regeneration is not simply "making new cells." It is spatial decision-making.

A concrete example shows how tightly this is managed. When a planarian is wounded, neoblasts across the body increase proliferation, not only near the cut. This systemic response is almost like a full-company alert. Yet the formation of a new head or tail is local. At the wound site, neoblast descendants gather and form a blastema-like outgrowth (planarians also use the word "blastema," though it differs from the salamander structure). Those cells then differentiate into the tissues appropriate for that location. If the wound is at an anterior-facing edge, the molecular map biases the outcome toward head structures; if it is posterior-facing, tail structures emerge. The same reserve is mobilized, but the instructions differ.

This distributed reserve has another consequence that is easy to overlook: it changes what injury *means*. In a salamander limb, the blastema must be organized at the stump and supplied with nerve-dependent growth signals. In planarians, the primary challenge is not to create a regeneration organ but to manage an organism-wide stem cell system so that it can be redirected without losing stability. The planarian does not need to persuade mature muscle to become youthful again in the same way; it already has youth on tap. The price is perpetual vigilance: a body that is always, quietly, in the business of rebuilding itself.

There is a human emotional resonance to this. Many of our hopes for regenerative medicine hinge on the idea of adding what we lack: more stem cells, more potency, more plasticity. Planarians suggest a different framing. Potent cells are only half the story. An immortal reserve is useful only if it is paired with reliable governance: patterning signals that prevent errors, genome-protective mechanisms that prevent corruption, and tissue environments that make differentiation an attractive exit rather than a reluctant surrender.

A planarian fragment regrowing a head is a spectacle. Underneath that spectacle is something less visible but more radical: a small animal living with a permanent population of cells that, in another context, would be considered dangerous, and yet using them not as a last resort but as an everyday tool. The worm glides along, unhurried, carrying within its soft body a distributed, pluripotent workforce and the molecular rules that keep that workforce building the same animal, again and again, without forgetting what "same" means.

2.3 Turning Back Time

The adult body is a museum of decisions. A muscle cell is not merely a cell that happens to be in muscle; it is a cell that has spent months or years reinforcing the same identity, tightening the screws on gene programs that say *contract, align*, do not divide*. Bone cells embed themselves in mineral; cartilage cells live in a gelatinous matrix they largely made themselves; skin cells stack into layers and then die on schedule. Specialization is what makes complex animals possible. It is also what makes regeneration such a scandal.

Planarians keep a supply of cells that never fully committed. Salamanders, for the most part, do not. When a salamander loses a limb, it cannot simply dispatch an all-purpose stem cell army from every corner of the body. It has to improvise locally. Some of that improvisation uses resident stem cells where they exist, and some of it depends on a stranger maneuver: mature cells near the injury edge loosen their identity and return to a more flexible state. That reversal is called *dedifferentiation*. The word is long because the idea is hard: to dedifferentiate is to walk backward down a path that development usually treats as one-way.

A useful image, proposed by the developmental biologist C. H. Waddington in the mid-twentieth century, turns cell fate into a landscape. Imagine a ball rolling down a hillside that has branching valleys. Near the top, the ball can take many routes. Farther down, the valleys deepen and the ball is guided into a specific groove: muscle, neuron, cartilage, blood. Differentiation is the roll downhill; dedifferentiation is the improbable push back up, out of a committed valley toward a higher, more plastic terrain.

Waddington-style Landscape

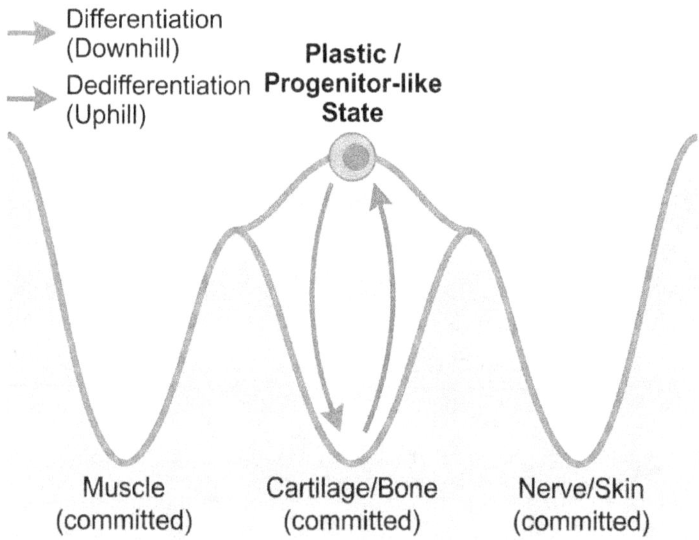

The landscape is a metaphor: valleys represent stable gene-expression programs. "Pushing uphill" (red) corresponds to molecular work that reopens closed DNA regions and reactivates growth programs, reversing differentiation (blue).

Metaphors become dangerous when they sound too tidy, so it helps to say what "uphill" means in molecular terms. Inside the nucleus,

DNA is wrapped around proteins to form chromatin, a packaging system that can be loose and accessible or tight and silent. When a cell differentiates, it does not only turn certain genes on; it locks many others away. Some of those locks are chemical marks on DNA or on the histone proteins around which DNA is wound. These marks do not change the genetic code, but they change how readable it is. That is why biologists call them *epigenetic* marks: "above" genetics. Differentiation is, in part, the gradual tightening of epigenetic constraints.

Dedifferentiation requires loosening those constraints without erasing the cell's entire identity in a catastrophic way. The salamander cannot afford to create a mass of cells that behave like a confused embryo. It needs cells that are youthful enough to proliferate, but disciplined enough to rebuild a limb rather than a tumor. The remarkable thing about salamanders is not that they can make cells plastic. It is that they can make them plastic *locally*, *temporarily*, and *under supervision.*

One of the first clues that mature tissues could reverse course came from older microscopy studies of newt limbs. After amputation, cells near the stump edge showed changes that looked like a retreat from specialization: muscle fibers broke down; connective tissue cells changed shape; cartilage-associated cells became more migratory. These observations were suggestive but not definitive. Seeing a cell "look younger" is not the same as showing that it re-entered the cell cycle and produced new tissues.

Modern lineage tracing made the question sharper. In axolotls, carefully labeling particular tissue types and following their descendants revealed that the blastema is not a uniform crowd of totally reset cells. As Kragl and Knapp and colleagues showed in 2009, many cells carry a memory of their tissue origin. They proliferate and contribute, but largely to their own lineages. The blastema is therefore a mosaic: a shared worksite assembled from multiple sources, not a single pool of identical raw material.

At first glance, this seems to demote dedifferentiation. If cells keep their lineage, how much "backward travel" is really occurring? The answer is that dedifferentiation is not all-or-nothing. A muscle cell can move "uphill" just enough to divide again and generate muscle progenitors, without becoming capable of turning into neuron or gut. A connective tissue cell can become proliferative and migratory without losing the basic identity that makes it connective tissue. In other words, dedifferentiation can mean returning to a progenitor-like state *within a lineage*. That is still extraordinary. In mammals, mature muscle fibers, for example, are famously reluctant to re-enter the cell cycle. Our muscle repair relies heavily on resident stem cells (satellite cells) rather than mature fibers reversing their fate.

Even among salamanders, the way "uphill" happens can differ. Sandoval-Guzmán, Wang, and colleagues (2014) compared muscle regeneration strategies across two salamander species and found a striking divergence. In some newts, mature muscle fibers can fragment and give rise to proliferative mononuclear cells that contribute to regenerated muscle, a process that looks like true dedifferentiation of mature tissue. In axolotls, by contrast, muscle regeneration relies more heavily on Pax7-positive satellite cells, a resident stem cell population more reminiscent of mammalian muscle repair. Two closely related animals, two workable solutions: one leans into cellular alchemy, the other leans into a quieter reserve.

This variety matters because it prevents a too-simple moral. Salamander regeneration is not a single recipe. It is a family of strategies that converge on the same result. Dedifferentiation is one powerful tool in that family, but it is not the only one, and it can be deployed in different tissues and different species to different degrees.

The deeper mystery becomes: what tells a mature cell that it is allowed to climb?

Part of the answer lies in the signals that keep the construction site active: the specialized wound epidermis and its apical epithelial cap, the remodeled extracellular matrix, and the immune environment. Another part lies in a requirement that has an almost cruel elegance: nerves.

The nerve dependence of limb regeneration is not simply a matter of sensation; nerves deliver molecular permission slips. When a limb is denervated, the wound can close, but the blastema falters. At the cellular level, the proliferative engine sputters. At the molecular level, key growth-promoting factors diminish. Jeremy Brockes and colleagues helped clarify one mechanism in 2007 by identifying nAG, a secreted protein of the anterior gradient family, whose expression is tied to nerve presence and appears in both nerve and wound epidermis during regeneration. When nAG was provided locally, it could partially rescue regeneration in denervated contexts. That kind of result changes nerve dependence from a vague requirement into a mechanistic link: nerves help sustain the signal environment that keeps cells in a growth-competent, plastic state.

Plasticity, however, is not enough. A limb is not a generic outgrowth; it has a specific architecture. If you cut a limb at the forearm, the animal regenerates the hand and wrist, not an entire limb starting from the shoulder. That implies that cells in the blastema carry information about *position*. They know where they are along the proximodistal axis (shoulder-to-fingertip), and they know what is missing.

A classic way to appreciate positional identity is through the effects of retinoic acid (RA), a derivative of vitamin A that acts as a powerful signaling molecule during development. In salamander limb regeneration, RA can reprogram patterning: at sufficiently high levels, it can "proximalize" the blastema, coaxing cells toward more proximal identities and sometimes producing duplications. The existence of this effect is both thrilling and alarming. Thrilling because it shows positional identity is not fixed like a serial number stamped in metal; it is a program that can be rewritten. Alarming because it reveals how easily the program can be miswritten.

Brockes and colleagues have long associated positional identity with cell-surface determinants such as Prod1, a protein whose expression varies along the limb axis and responds to RA. More recently, other RA-responsive determinants have been identified, including molecules such as Tig1/Rarres1 that can influence positional programs at the cell surface. The conceptual point is that positional memory is not an abstract coordinate; it is embodied in molecules that cells display and interpret, like badges in a workplace that determine who is allowed to enter which rooms.

Regenerating tissue also controls RA signaling not only by making RA but by destroying it. Enzymes such as CYP26B1 degrade RA, reducing signaling. Inhibiting CYP26B1 can raise RA signaling and shift patterning outcomes, again pushing cells toward proximal identities. This is an elegant form of regulation: instead of relying solely on a faucet (synthesis), the system also uses a drain (degradation). The level in the sink is what matters.

All of this intersects with dedifferentiation in a way that is easy to miss. To reverse a mature cell's fate safely, the body must do two contradictory things at once. It must reopen growth programs and loosen chromatin constraints, but it must also keep cells anchored to the correct positional plan. A cell that becomes plastic without positional discipline is not a regenerative solution; it is a developmental problem.

That is why the blastema is so often described as a place where cells are both freer and more constrained than usual. Freer in that they divide, migrate, and temporarily abandon some features of maturity. More constrained in that they are embedded in a signaling-rich environment that continually reminds them what they are allowed to become. The construction site is permissive; the permit is specific.

The immune system adds another layer of specificity through timing. In adult salamanders, early macrophage activity is required for successful regeneration; depleting macrophages early leads to fibrosis and failure. One way to interpret this is that macrophages participate in clearing damaged tissue and shaping the extracellular matrix so that dedifferentiating and progenitor cells can move and proliferate. Another is that macrophages help resolve inflammation in a way that avoids locking fibroblasts into scar-producing behavior. In mammals, chronic or mis-timed inflammation can trap tissues in a loop of repair that never transitions to reconstruction. Salamanders appear to handle inflammation like a controlled burn: necessary, dangerous, and tightly managed.

If the story stopped here, dedifferentiation would be merely a technical curiosity. Its human significance comes from the fact that we have learned to induce something like it artificially. The most famous example is the creation of induced pluripotent stem cells, achieved by introducing a set of transcription factors that can push adult cells back toward an embryonic-like state. That work, associated with Shinya Yamanaka's lab (2006), demonstrated that adult cell identity is reversible in principle. Yet the laboratory method often carries risks: genomic stress, incomplete reprogramming, tumor formation if cells remain too plastic for too long. The salamander performs a version of reprogramming in living tissue with far fewer catastrophic outcomes, suggesting that the missing ingredients in humans may be less about possibility and more about governance: when, where, how far, and under what constraints cells are allowed to climb.

One concrete example of "how far" being the key comes from the very heterogeneity of the salamander blastema. If blastema cells were truly pluripotent in a broad sense, one might expect widespread lineage crossing. Instead, much of the regeneration proceeds with lineage restriction intact. This may be part of the safety system. A muscle-derived progenitor that remains muscle-biased is less likely to misbuild cartilage in the wrong place. A connective tissue progenitor that carries positional identity can help instruct other tissues without itself becoming something unrelated. Dedifferentiation, in this framing, is not the erasure of identity but the temporary relaxation of adulthood's strictest prohibitions: *you may divide again*; *you may move*; *you may respond to developmental signals again*.

A final, sobering point is that "turning back time" is not free. In mammals, strong anti-cancer defenses, long lifespans, and complex immune systems are tied to how tightly we restrict cell plasticity. If salamanders can relax those restrictions and still avoid disaster, they are doing so with compensations we do not yet fully understand. Some of those compensations may be genetic and species-specific; some may lie in the microenvironmental choreography of regeneration; some may be in tumor suppression strategies that differ from our own.

In the salamander stump, dedifferentiation looks less like rebellion and more like a sanctioned emergency protocol. Mature cells loosen their grip on a single identity, but they do so inside a tissue that has been converted into a tightly policed construction zone. Signals from the wound epidermis, nerves, immune cells, and positional gradients keep the reversal purposeful. When the limb is rebuilt, the permissions are revoked. Cells settle back into their valleys, chromatin tightens, proliferation subsides, and the rebuilt structures return to the quiet, disciplined labor of adult tissue—strong, specialized, and, until the next injury, content to stay exactly what they are.

Chapter 3

Maps Made of Molecules

Imagine cutting a map in half and watching it redraw the missing coastline in perfect detail; biology achieves this through invisible chemical coordinate systems that permeate living tissue. This chapter uncovers the sophisticated molecular logic-from concentration gradients to mathematical feedback loops-that tells a cluster of cells whether to build a shoulder or a thumb, and exactly when to down tools.

3.1 Reading the Invisible Map

A severed salamander limb does something that feels almost insolent. Within weeks it can replace bone, muscle, tendons, nerves, skin, and blood vessels, and it does so with a quiet confidence about proportion and orientation. The new structure is not a generic spike of tissue; it knows where the elbow belongs, how long the forearm should be, where the wrist should end, and what counts as a fingertip. Somewhere inside that wound, ordinary-looking cells act as if they have access to a hidden atlas.

The puzzle is not simply that cells divide and specialize. It is that they do so *in the right places*. A cell sitting near the shoulder must behave differently from a cell sitting near the wrist, even if they share the same DNA and were recently neighbors. Humans encounter a faint analogy when a friend gives directions in a city: "two blocks north, then turn west." The words only make sense because there is a coordinate system built into the streets. Living tissues need their own coordinate systems, but they cannot depend on streets, compasses, or any device that seems remotely like a map.

Lewis Wolpert, a developmental biologist with a gift for clarifying foggy ideas, gave the field a phrase that still guides the conversation: *positional information.* In his 1969 paper, he argued that making a body part requires two logically distinct operations. First, a tissue must establish a coordinate-like system so that cells can possess a "positional value"—a kind of molecular answer to the question "where am I?" Second, cells must *interpret* that value, converting it into a decision: become cartilage, or epidermis, or tendon; begin dividing, or stop. Separating those two steps sounds fussy, but it is as practical as distinguishing a thermometer from a doctor's judgment. A thermometer gives a number; medicine decides what to do about it.

How could a tissue hand out positional values? Wolpert offered a model so simple it has become a rite of passage in biology courses: the French Flag model. Imagine a strip of tissue, and imagine a single chemical signal—a *morphogen*, meaning a "form-maker"—produced at one end and spreading through the strip. Cells near the source experience a high concentration; cells farther away, a lower one. Now impose two thresholds: above the higher threshold, cells adopt fate A; between thresholds, fate B; below the lower threshold, fate C. A smooth chemical gradient is read out as sharp bands of identity. The model is not a literal claim that embryos carry tricolours inside them. It is an argument that a graded signal, if interpreted with thresholds, can turn continuous space into discrete anatomy.

At first glance this feels like magic by arithmetic: a cell measures a number and picks a label. But the deeper idea is that chemical concentration can *stand in* for distance. A cell cannot directly sense "I am 3 millimeters from the shoulder," yet it can sense "I am bathing in 17 nanomolar of signal X." If the tissue reliably translates physical distance into chemical concentration, then chemistry becomes geometry.

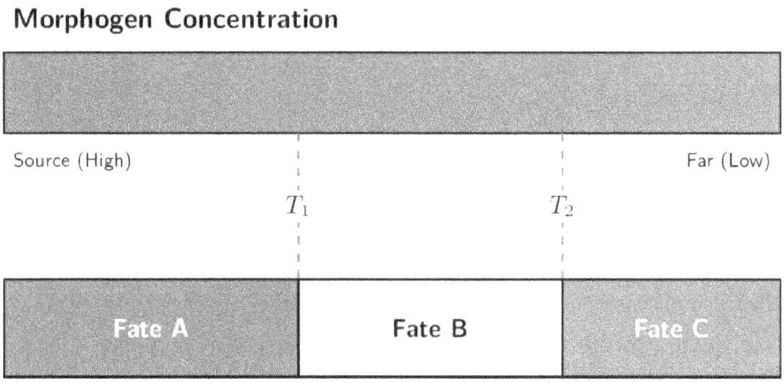

French Flag Model: Cells interpret a continuous morphogen gradient using concentration thresholds (T_1, T_2) to induce discrete cell fates.

The French Flag picture is the invitation, not the destination. Real tissues must solve complications that a chalkboard can ignore. The gradient must be *made* in three-dimensional, wet, irregular living material. It must be stable enough to guide pattern but flexible enough to repair after injury. It must be read in the midst of noise: cells are jostled, receptors fluctuate, molecules degrade, blood flow and extracellular matrix can distort movement. And in regeneration, it must often be rebuilt in adulthood, not merely laid down once in an embryo.

Start with the physics. If a group of cells secretes a signaling molecule, that molecule moves outward by diffusion—the same basic process that spreads perfume in a room. Diffusion alone would eventually make the concentration uniform everywhere, erasing positional differences. To preserve a gradient, the tissue needs counterforces: degradation, uptake by receptors, binding to extracellular scaffolds, or transport that is biased. The simplest stabilizing arrangement is a source and a sink: production at one end and removal at the other, or production in one region and degradation throughout. Under those conditions the molecule can settle into a steady state where concentration falls with distance. Even without mathematics, the intuition is familiar: a candle makes a bright region near the flame and a dim region far away because photons are "removed" by geometry and absorption; a morphogen makes a high region near its source and a low region far away because molecules are removed by decay and capture.

Cells then need a way to measure concentration. Typically they do this with receptors on their surface: proteins that bind the signaling molecule and trigger an internal biochemical cascade. The cascade often ends in the nucleus, altering the activity of genes. "Gene" here means a DNA sequence that can be transcribed into RNA and often translated into protein. When a morphogen pathway activates a gene, it changes what the cell makes, which changes what the cell does. In the French Flag model, different genes would turn on at different concentration ranges, locking in distinct cell fates.

This begins to sound like a tidy thermostat: if the signal is above threshold, flip the switch. But biological thresholds are rarely single switches; they are created by networks. One common trick is mutual repression. Suppose gene A and gene B inhibit each other. If the morphogen slightly favors A, A will suppress B, and the cell will end up strongly A-like. If the morphogen favors B, the opposite happens. The result is a sharpening of boundaries: a narrow region of intermediate morphogen becomes a crisp border between two gene-expression states. Another trick is the "repressor of a repressor" design, in which the morphogen turns off a brake, thereby allowing a target gene to rise. These designs matter because they turn a somewhat unreliable measurement into a stable decision.

Noise is not a minor detail; it is the central engineering challenge. Individual cells do not count molecules perfectly. A receptor might be abundant in one cell and sparse in another. Internal molecules bump into each other randomly, and gene expression comes in bursts. Yet embryos and regenerating tissues are often extraordinarily reproducible. If a positional system were as fragile as the French Flag cartoon might suggest, development would be a lottery.

One solution is time. A cell can integrate the signal over minutes or hours rather than reacting instantly to momentary fluctuations. Another is feedback. Many signaling pathways induce their own inhibitors: a cell exposed to high morphogen may respond by making more receptor (which can increase uptake and effectively dampen spread) or by producing an inhibitory protein that reduces pathway output. This can create *adaptation*, where the pathway output initially rises but then settles even if the external signal stays high. Adaptation sounds like sabotage, but it can prevent runaway responses and help the system focus on changes rather than absolute levels. In some contexts, what matters is not "how much morphogen is present" but "how long have I been above a certain level" or "how rapidly did it rise." A gradient then becomes not a static ruler but a timed instruction.

Another solution is redundancy. Tissues rarely rely on a single chemical to encode position. They combine multiple signals, often arranged along different axes. A limb, for instance, must distinguish shoulder-to-fingertip (proximal–distal), thumb-to-little-finger (anterior–posterior), and back-of-hand to palm (dorsal–ventral). A single one-dimensional gradient cannot label a three-dimensional object. Instead, a cell reads combinations: high signal X plus low signal Y plus intermediate signal Z. The logic is closer to GPS than to a street sign: position is inferred from multiple beacons.

Regeneration adds an additional twist. In an embryo, positional information is laid down as tissues form; in a wound, tissue must reconstruct the coordinate system while simultaneously repairing. In salamander limb regeneration, for example, the stump forms a *blastema*, a mound of proliferating cells that will rebuild the missing parts. Those cells are not identical blank slates. They come with histories: some are derived from muscle, some from cartilage, some from connective tissue, and their lineages constrain what they can become. Yet they still must agree on large-scale geometry: how much to rebuild and where.

Here a historical anecdote helps, because it reveals how biologists first learned that cells carry positional memories at all. In the middle of the twentieth century, long before gene sequencing, researchers performed transplant experiments on amphibian limbs. They would rotate a piece of tissue, swap it from one side to another, and watch what grew. When tissues with different positional identities were juxtaposed, regeneration could produce extra structures: additional digits or mirror-image duplications. The tissue behaved as if it were trying to resolve a contradiction: two neighboring regions claiming incompatible coordinates. The solution was not compromise but *intercalation*—filling in missing positional values between them, sometimes generating supernumerary parts in the process. These experiments were messy, alive with surgical improvisation, yet they delivered a crisp conceptual point: cells are not merely responding to the wound; they are comparing positional identities and reconstructing a complete pattern.

The French Flag model supplies one way to understand how such identities could be assigned. A gradient could label the stump, and blastema cells could read it, choosing fates appropriate to their measured concentration. But transplantation results also hint that positional information is not only about present exposure; it can be stored. "Stored" here does not mean a tiny map tucked inside the cytoplasm. It means that a cell's gene regulatory network—the web of genes turning each other on and off—can settle into stable states that persist even when the original signal changes. In other words, a transient gradient can leave a lasting imprint.

To see how a transient cue can become a stable memory, imagine a gene that, once turned on, helps keep itself on. Biologists call this positive feedback. If the morphogen nudges that gene above a certain activity, the gene's own feedback can lock it into the "on" state, even after the morphogen falls. Neighboring cells that never crossed that tipping point stay "off." A smooth gradient can therefore be converted into a mosaic of stable states: a memory of position written into gene expression.

There is also a practical reason why biological maps are usually not a single clean gradient from one edge of a tissue to the other. Tissues grow, shift, and remodel. A gradient that is perfect for a small limb bud may become stretched and diluted as the limb grows. Patterning systems often require *scaling*: if the tissue doubles in size, positional readouts must adjust so that proportions stay reasonable. Otherwise an organism might build too many digits or place joints at the wrong relative positions. Scaling can be achieved by feedback that adjusts how far a signal spreads, by regulating receptor abundance, by altering degradation rates, or by deploying additional molecules that act like expanders and contracters of the gradient's reach. The details vary across organisms and organs, but the recurring theme is that positional information is a *control system*, not a passive stain.

A concrete example makes the abstract machinery feel less like a metaphor. In many developing tissues, a signaling molecule is produced in a localized region and forms a gradient across a field of cells. Those cells respond by activating distinct sets of genes at different distances from the source. The readout is not a single threshold but a set of nested domains. The most sensitive targets respond far away; the less sensitive ones respond only near the source. Sensitivity can be tuned by the affinity of receptors, the strength of internal signaling, and the architecture of gene regulation. This is how a single chemical can have multiple meanings. To a cell near the source, it says "become type A"; to a cell at mid-range, "become type B"; and to a far cell, "stay type C." The same molecule is not delivering three messages; it is delivering one graded influence that different gene circuits decode differently.

The idea of decoding is the real heart of the invisible map. A morphogen gradient is not a set of instructions written in chemical ink; it is a field that biases probabilities and flips switches inside cells. If the field is the map, the gene regulatory network is the reader. Two tissues can be bathed in the same gradient and interpret it differently because their internal circuits differ. Even within one tissue, the same gradient can be interpreted in stages: early on, it may set broad territories; later, it may refine boundaries or control growth. This layered interpretation is part of why living patterning can appear both robust and adaptable.

That adaptability also helps explain why mammals, including humans, generally fail at large-scale regeneration. It is not simply that we lack the molecules. Many of the same signaling pathways exist in us, and our cells can respond to gradients during development. The problem is that adult mammalian tissues have a strong tendency to treat major injury as an emergency demanding rapid closure. The biochemical environment becomes dominated by inflammation and scarring. Scars are useful: they seal wounds quickly and reduce infection. But scarring can scramble the conditions needed for precise positional mapping. A gradient needs a permissive medium: cells that can move, signals that can spread, extracellular matrix that can be remodeled, and time for pattern to be established and interpreted. A fast, fibrotic patch is closer to a welded plate than to a rebuilding blueprint.

There is a final, more intimate reason this molecular cartography matters. It reframes the body as a communication system. When a limb is intact, cells continuously exchange information about neighborhood, tension, density, and chemical milieu. When a limb is amputated, the most astonishing feat is not that cells grow; it is that they coordinate growth into a coherent form. Morphogen gradients are one of the few mechanisms known that can tell a cell something about its location without giving it a ruler. They make distance legible.

A person watching a salamander regenerate may feel that the animal is "remembering" its limb. In a sense, it is. But the memory is not a ghostly picture of a hand. It is a set of molecular relationships that allow cells to recreate coordinates and then to act on them. The map is invisible because it is not drawn in lines; it is drawn in concentrations, thresholds, feedback loops, and the patient arithmetic of cells comparing chemical levels to internal states. When that arithmetic works, anatomy emerges with the quiet inevitability of a well-read address.

3.2 The Mathematics of Fingers and Toes

A chemical gradient can tell a cell whether it lives nearer the shoulder or nearer the wrist. That kind of information is already a marvel: a tissue that has been cut learns how far it is from a reference point without a ruler. Yet gradients alone, even perfectly read, would leave anatomy oddly vague. They are excellent at drawing borders and territories: here is proximal, there is distal; here is dorsal, there is ventral. But fingers and toes are not territories. They are repeated structures, a set of separated pillars rather than a single block. To make digits, a limb must transform a smooth field of "hand" into a pattern of alternating "finger" and "space between fingers." A gradient is a sentence; digits require punctuation.

The surprising claim is that the punctuation can arise from chemistry doing something that, at first glance, looks like a mathematical trick. A developing limb can begin as a relatively uniform sheet of cells, with no obvious periodic structure, and nevertheless produce a regular array of condensations that will become bones. This is the sort of pattern that makes people reach for blueprints, as if each future finger were marked in advance. Alan Turing suggested in 1952 that a blueprint is not always necessary. Under the right conditions, a chemical system can *make* its own pattern.

Turing is best known for codebreaking and the foundations of computing, which makes his biological paper feel like a message in the wrong bottle. It appeared when the idea of gene regulation was in its infancy and when molecules were not yet the everyday language of development. His title, "The Chemical Basis of Morphogenesis," was almost audacious: he proposed that an initially uniform tissue could become unstable and spontaneously split itself into spots and stripes because two chemicals, diffusing at different rates, react with one another. Today the phrase "Turing pattern" is attached to an entire family of such mechanisms, but the core intuition is compact enough to explain at a dinner party, provided there is a napkin and enough patience.

Imagine two substances produced by cells in a tissue. Call one an *activator*: it encourages its own production and also stimulates production of the second substance. Call the second an *inhibitor*: it suppresses the activator. If both stayed put, the system would settle into a uniform balance. The twist is diffusion. Suppose the inhibitor spreads faster than the activator. Then a small accidental bump of activator in one location has two effects. Locally it reinforces itself, because it is an activator. But it also creates inhibitor, and the inhibitor runs ahead, diffusing outward and suppressing activator in neighboring regions. The result is a patchwork: a peak where activator wins, surrounded by troughs where inhibitor keeps activator down. Peaks repel each other, in a sense, because the inhibitor halo makes it hard for a nearby peak to form. From an initially smooth field, a set of evenly spaced maxima can appear, as if the tissue had decided on a rhythm.

That rhythm has a physical consequence in the limb. Early in limb development, the future handplate is a pad of tissue. Later, narrow regions of cells condense into cartilage primordia—early skeletal elements—while neighboring cells remain less condensed or follow other fates. The spacing between these condensations becomes the spacing between digits. The regions between digits do not simply fail to become fingers; in many species they actively undergo programmed cell death, a normal and sculpting kind of cellular suicide. The final hand is therefore a pattern etched by both construction and subtraction.

The attraction of a Turing-like mechanism is that it naturally produces repeated elements with characteristic spacing, and it can do so without a separate "digit counter." It also explains why digits are so sensitive to the size and growth of the tissue. If the domain is larger, more peaks can fit; if it is smaller, fewer. Anyone who has watched a photograph of limb buds across species has seen the same basic ingredients—buds, pads, condensations—reused at different scales, yielding different numbers and thicknesses of digits. A self-organizing mechanism that responds to domain size is an economical way to vary outcome without rewriting the entire recipe.

The obvious objection is that biology is not two anonymous chemicals in a beaker. Real tissues have dozens of interacting signals, and limbs are patterned along multiple axes. Nevertheless, one of the striking developments of the last decade is that plausible molecular candidates for an activator—inhibitor network in digit formation have emerged. Work around 2014 proposed that a network involving BMP signaling, WNT signaling, and the transcription factor SOX9 can behave like a Turing system: SOX9 marks chondrogenic condensations (the beginnings of digits), while BMP and WNT participate in feedback loops that can amplify local differences and impose spacing. No single molecule wears the badge "activator" in all contexts, because molecules are promiscuous—BMP can promote differentiation in one setting and restrain it in another, depending on what other signals are present—but the *logic* of local positive feedback coupled to longer-range inhibition is what matters.

This is where the earlier notion of positional information quietly returns, not as repetition but as a necessary partner. Self-organization is powerful, but it is also indifferent. A pure reaction-diffusion system can create stripes oriented in any direction. It can form spots where stripes would be disastrous. It can place a peak inconveniently close to a boundary unless the boundary itself constrains it. A limb, by contrast, must align its pattern with the long axis of the handplate and must coordinate digit formation with overall growth and with identity: a thumb is not simply "digit number one from a random edge." Large-scale gradients and boundary cues can bias the self-organizing mechanism so that the same local chemistry produces the right global architecture.

One way to picture this is to imagine a field of grass in which wind creates ripples. The ripples are self-organized, but the shape of the valley and the presence of trees influence their direction and spacing. In a limb, broad gradients and anatomical boundaries can act like the valley and trees. They set conditions—how readily cells respond to BMP, where WNT is strong or weak, where growth is fastest—that tune the reaction-diffusion system. The mathematics does not replace genetics or anatomy; it describes a particular kind of collaboration between them.

A concrete example of that collaboration lies in a long-running debate about Sonic hedgehog (SHH), a signaling molecule famous for its role in patterning the anterior-posterior axis of the limb, among many other roles. For years, SHH was often presented as a classic morphogen for digit identity: a molecule emanating from the posterior side of the limb bud, forming a gradient, with higher exposure specifying more posterior digits. This story fits comfortably with the "French Flag idea": different thresholds yield different digit types. Yet more recent experimental work has complicated that comfortable picture. In mice, evidence suggests SHH may act less like a long-range "you are here" ruler for each digit and more like an early trigger that enables digit formation through downstream processes, including survival and outgrowth, with identity emerging from a combination of timing, gene network dynamics, and interactions with other signals. When researchers bypass certain survival constraints, brief pulses of SHH can suffice to set in motion the formation of multiple digits. The chemistry still matters enormously, but the mechanism looks less like a static gradient read by thresholds and more like a temporal instruction that sets parameters for a self-organizing system.

This is not a mere argument about definitions. Whether SHH is "a morphogen" in the strict sense changes what kind of explanation one looks for. A strict morphogen story implies that the spatial profile of SHH concentration, at the time cells make fate decisions, is the central code. A trigger-and-network story implies that pathway architecture—feedback, adaptation, and gene regulatory states—can transform a transient signal into a lasting pattern, and that periodic digit spacing may be generated by other interacting molecules once the system is primed. Both stories can be partly true in different species and at different stages, which is precisely why limbs remain such an instructive arena. They force biology to admit that "concentration" is rarely the whole message.

Turing himself would likely have enjoyed this messiness. His 1952 paper did not insist on any particular molecule; it offered a class of mechanisms. It also arrived with a kind of intellectual loneliness. The biology of the time did not have the experimental tools to test his proposals directly, and many readers found the mathematics forbidding. The historical irony is that the modern era, obsessed with genetic parts lists, has returned to his argument because the parts lists alone did not yield an intuitive blueprint for form. Genes can tell cells what proteins to make, but they do not, by themselves, explain why a uniform tissue becomes a striped one. Pattern formation is not only about ingredients; it is about dynamics.

Dynamics, here, means that the system's behavior depends on time and on interactions. In a reaction-diffusion mechanism, the crucial events are instabilities: moments when a uniform state becomes fragile and a small fluctuation gets amplified rather than damped. Biologists sometimes describe this as the tissue "breaking symmetry," but that phrase can sound grander than the underlying reality. Cells are noisy chemical factories. Somewhere, a cluster of cells will by chance have a touch more of some signal. The question is what the system does with that accident. Most of the time, biology is designed to correct accidents. A Turing regime is special because it can convert a small accident into a stable feature.

Limb tissue also grows while it patterns, which adds a further layer of mathematics. A pattern that forms on a growing domain can be stretched, duplicated, or re-spaced. If a reaction-diffusion system is operating while the handplate expands, new peaks can appear between old ones, increasing digit number or altering spacing. This is one reason why the timing of growth matters so much. Mutations or experimental perturbations that slightly change growth rate, or the duration of a signaling window, can have outsized effects on pattern outcome. The system is not writing digits into a static canvas; it is drawing on a canvas that is enlarging under the pen.

Regeneration brings these ideas into even sharper focus because it asks the same machinery to operate under adult conditions. When a salamander regrows a limb, the blastema must produce a hand with the correct number of digits and spacing appropriate to the animal. It must do this while nerves re-enter, blood supply is restored, and tissues are re-differentiating. The adult body is not a clean embryonic environment, yet the pattern emerges with a reliability that is hard to attribute to a rigid pre-plan. A self-organizing system, tuned by positional cues, is one of the few explanations that can plausibly be both robust and flexible.

There is, however, a subtle emotional sting to this elegance. A mechanism that can reliably generate fingers sounds like exactly the sort of thing humans would want to borrow for regenerative medicine. Why not simply switch it back on after an amputation? The uncomfortable answer is that a digit-making system is inseparable from a growth-making system. If the activator—inhibitor logic that spaces digits is coupled to proliferation and survival, then reactivating it in adult human tissue risks the wrong kind of growth. Embryos can afford exuberant proliferation because they are building an organism that is already committed to growth; adult tissues must constantly balance repair against the ever-present danger of tumors. The same mathematics that creates a beautiful pattern can, in the wrong context, create a pathology.

Even in embryos, the system must be constrained. A reaction-diffusion mechanism can in principle generate endless patterns, but limbs are finite. Boundaries matter: the edges of the handplate, the distribution of extracellular matrix, the presence of neighboring tissues, the geometry of the domain. Biology uses these constraints as part of the computation. A boundary can act like a sink that absorbs a molecule, or like a source that produces it, or like a barrier that blocks diffusion. Each of these changes what patterns can form. An edge is not merely the place where the tissue ends; it is a line of instruction.

So a hand is built from two kinds of information woven together. One kind is coordinate-like, giving cells a sense of large-scale position and orientation. The other kind is self-organizing, able to produce repeated elements with the right spacing from local interactions. The common mistake is to imagine that one of these must be the "real" explanation. In practice, the elegance of limbs comes from the way a broad cue can guide a local instability so that the resulting pattern is not merely regular but anatomically useful.

The appeal of Turing's idea, then, is not that it reduces life to equations. It is that it gives a language for something biology does all the time: it turns small molecular differences into large, stable structures. Fingers and toes are easy to take for granted because they are so familiar. Yet their existence depends on a tissue's willingness to amplify a whisper into a ridge, to let chemistry create spacing where no carpenter has marked the wood, and to do so while the entire structure is growing and reshaping itself. When the peaks of a molecular pattern become the scaffolds of future bones, mathematics stops being an abstract description and becomes a form of anatomy you can hold in your hand.

3.3 The Molecular Brake Pedal

Regeneration is often described with the language of miracles: lost parts reappear, wounds close without scars, function returns as if nothing happened. The word hides a darker truth. Growth is not inherently benevolent. A tissue that knows how to expand quickly is only a few regulatory mistakes away from behaving like a tumor. Every animal capable of dramatic repair must solve a problem that is almost philosophical: how to make growth conditional, finite, and exquisitely measured. A salamander's regrown limb does not stop because it is tired; it stops because it receives, and obeys, signals that say *enough.*

The earlier puzzles of regeneration can be phrased as questions of direction and pattern. How does a limb know which end is shoulder and which is fingertip? How does it turn a smooth field into a repeated structure like digits? The stopping problem is different. It is about restraint. It asks how a living system translates an abstract goal—restore normal size and shape—into a concrete sequence of molecular events that end in quiescence, the cellular state of not dividing.

A first hint comes from an organ that humans *do* regenerate well: the liver. If part of the liver is removed, the remaining tissue grows until the original functional mass is restored. This is sometimes reported as "the liver regenerates," which is true in spirit but misleading in detail. The liver does not regrow a missing lobe with the same anatomical choreography seen in a salamander limb. Instead, the existing liver enlarges and remodels until overall capacity is sufficient. Even so, it demonstrates something crucial: adult mammalian tissues are capable of robust, coordinated growth, and they are capable of stopping on time. The existence of that stop is not an afterthought; it is the feature that makes liver regrowth safe.

In the twentieth century, surgeons and physiologists studying partial hepatectomy—surgical removal of part of the liver—were struck by the reproducibility of the response. The liver's mass could return with remarkable accuracy, and the dynamics depended on both biochemical signals and blood flow. It was a living lesson in feedback control: remove tissue, and the system accelerates proliferation; restore tissue, and proliferation shuts down. The question is what the tissue is measuring. It cannot count hepatocytes one by one. It must rely on proxies.

Cells have an abundance of proxies, and they come in two broad flavors. One is chemical: growth factors and morphogens, the same family of signals that can create gradients and patterns, can also stimulate or restrain proliferation. The other is physical: cells can sense crowding, stretch, stiffness of their surroundings, and the forces transmitted through their attachments to neighbors and to the extracellular matrix. The physical category is easy to underestimate because it lacks a tidy molecular name, but it is as real as a bruise.

The most prominent molecular bridge between physical sensation and growth control is the Hippo pathway. The name is memorable in the way biology sometimes allows itself: in fruit flies, mutations in components of this pathway produced tissue overgrowths, and the phenotype reminded researchers of a hippopotamus. Behind the whimsy sits a deeply conserved system. Versions of the pathway operate in animals from flies to humans, and a simplified description captures its essence. When cells are sparse, stretched, or otherwise in a pro-growth mechanical state, two transcriptional co-activators called YAP and TAZ are relatively active. "Co-activator" means they do not bind DNA on their own as classic transcription factors do, but they partner with DNA-binding proteins to turn genes on. YAP/TAZ activity pushes cells toward proliferation and survival. When cells become crowded and tightly packed, the core Hippo kinases switch on and chemically modify YAP/TAZ so they are kept out of the nucleus or marked for degradation. The nucleus, abruptly deprived of YAP/TAZ help, turns down growth-promoting genes. The tissue brakes.

This is satisfying because it makes intuitive sense. If cells are bumping into one another, there may be no room to expand. But the deeper importance is that it gives tissues a way to measure size without measuring size directly. A tissue at the right scale tends to have characteristic density and tension. A regrowing organ can therefore stop when mechanical conditions return to their baseline. The stop signal is not an external announcement; it is a local readout of geometry.

Mechanical feedback alone cannot explain everything, because tissues can be crowded and still need to grow, and tissues can be sparse and still need to stop. The Hippo pathway itself listens not only to cell density but also to multiple upstream cues: cell polarity (how cells know which side is "top" and "bottom"), the architecture of cell junctions, and signals from the extracellular matrix. It behaves less like a single brake lever and more like a braking system that reads several sensors and produces a combined decision.

Regeneration adds an extra layer: growth must stop not merely at the right *volume* but at the right *patterned shape*. A lumpy hand is not a hand. This is where the earlier mapping machinery—gradients and self-organized patterns—returns in a new guise. Signals that originally helped position and sculpt structures often also regulate growth rate. In development, morphogens are famous for patterning, but many also act as growth regulators. They can encourage proliferation in one concentration range and inhibit it in another, or they can control growth indirectly by repressing growth inhibitors. A tissue can therefore tie "where you are" to "how much you should divide," preventing a part from overgrowing relative to its neighbors.

A concrete illustration comes from Drosophila, the fruit fly, which has served as an intellectual wind tunnel for testing ideas about growth control. In the developing wing, the morphogen Decapentaplegic (Dpp) forms a gradient and patterns the tissue. For years, a tempting story was that cells might measure the *slope* of the gradient—how quickly concentration changes across space—and use that to regulate proliferation, ensuring even growth across the wing. The trouble with slope sensing is that it is technically demanding: it asks a single cell to compare itself with its neighbors and to compute a derivative in a noisy chemical environment. A different, more biologically plausible mechanism emerged in which Dpp's main growth effect can be mediated through repression of a gene called *brinker* (brk), which functions as a growth repressor. High Dpp keeps brk low; where brk rises, growth is restrained. In that logic, Dpp does not need to provide a precise "grow at this rate" command everywhere. It needs to keep a brake (brk) released in the right places, and the tissue's default programs can handle the rest. It is an elegant inversion: growth is controlled not by pressing the accelerator but by lifting the foot off the brake.

This inversion is common in biology, partly because repression is often more robust than activation. Turning something off can be achieved by multiple mechanisms—blocking transcription, destabilizing RNA, inhibiting proteins—so the system gains redundancy. Repressors can also be integrated easily into feedback loops. If growth begins to exceed what is appropriate, a repressor can rise and clamp it down. The point is not that biology prefers pessimism; it is that biological circuits are often built to fail safe.

Feedback loops also solve another problem that becomes painfully obvious once one thinks like an engineer: scaling. A gradient that patterns a tissue of one size must still pattern it if the tissue is larger or smaller, whether due to natural variation or because the system is regrowing after injury. If the gradient's reach were fixed, then a larger tissue would have an unpatterned region at the far end. Scaling requires mechanisms that adjust the effective range of signals to the size of the domain. Several models achieve this, one of which has a pleasingly paradoxical name: the "expander." In some proposed circuits, a molecule produced in low-signal regions increases the spread of the morphogen, while the morphogen suppresses the expander. This creates a self-adjusting system: if the tissue is too large, the expander has more room to accumulate, stretching the gradient; if the tissue is too small, the morphogen suppresses the expander more effectively, contracting the gradient. A pattern can keep its proportions even as the tissue grows.

The most interesting part of scaling mechanisms is that they blur boundaries between disciplines. A signal's reach is not merely a chemical property; it depends on tissue architecture and cell biology. Molecules can be internalized by receptors, trafficked through endosomes, recycled back to the surface, or transported through cellular extensions. These processes are the daily work of cells, usually discussed in textbooks under "membrane trafficking," yet they can determine how far a morphogen effectively travels and how a gradient evolves over time. A system can scale not because a molecule diffuses farther in a larger tissue, but because cells in a larger tissue collectively recycle and redistribute it differently.

If all of this sounds too elaborate for a creature that simply needs to regrow a tail, it is worth remembering what the tail represents: a large quantity of organized tissue. It contains spinal cord, muscle, bone or cartilage, blood vessels, skin, and nerves. Regenerating it means coordinating many cell types so that no one population gets ahead of the others. The stop signal must be shared across those populations, and it must be legible to each of them in their own biochemical dialect.

Some of the most persuasive evidence that mechanical braking matters in regeneration comes from the banal fact that tissues behave differently when their mechanical environment changes. Cells grown on soft gels behave differently from cells grown on stiff plastic. In living animals, stiffness changes with age, inflammation, and scarring. A scar is not just collagen laid down as a patch; it is a mechanical reconfiguration. A stiff fibrotic matrix can keep YAP/TAZ activity elevated and bias cells toward proliferation and altered differentiation. This is one reason scarring is not merely an aesthetic issue. It changes the decision landscape. A regenerative program trying to stop at the right time must do so in a tissue whose mechanical cues are shifting under its feet.

This brings us to a painful trade-off that shadows much of regeneration biology. Mammals, in general, have evolved potent tumor-suppression strategies and rapid wound closure. Those strategies involve inflammation, fibrotic sealing, and tight control of cell-cycle re-entry. They are excellent at preventing death from bleeding or infection, and they reduce the chance that a wounded area becomes a cradle for cancer. But they can interfere with the slow, pattern-sensitive rebuilding that full regeneration requires. In that light, the molecular brake pedal is not merely a device to stop regeneration; in many mammals it is engaged early, hard, and for good reasons.

Even in animals that regenerate well, the brake must be carefully calibrated. Too little braking yields overgrowth or malformed structures. Too much braking yields stumps. The elegance of systems like Hippo is that they can be gradual. They do not force an all-or-none stop; they allow tissues to taper growth as mechanical conditions shift. Chemical feedback circuits do something similar by changing signal levels smoothly and by altering sensitivity through receptor regulation and downstream inhibitors. Braking, in practice, is not a wall; it is a slope that becomes steeper as the target size is approached.

The most human way to appreciate this is to imagine what "correct size" means in the first place. A salamander's limb must match the other limb, not an abstract ideal. A liver must match the organism's metabolic needs, which depend on body size and on diet. A child's organs must grow in proportion as the child grows. The target is not a fixed number but a moving equilibrium. Biology therefore needs brakes that are not set to a single stopping distance. It needs brakes that sense context.

Mechanical feedback provides context in a particularly intimate way because it is local and immediate. A cell does not need to know the organism's weight; it needs to know whether it is stretched, whether its neighbors are close, whether its attachments are under tension. Chemical feedback provides context in a complementary way because it integrates information over longer distances. A gradient can tell a cell it is near an edge; a growth factor can report injury; an inhibitory signal can spread from a region that has already reached maturity. A regrowing tissue stops not because one master signal declares completion, but because multiple sources of context converge to make proliferation increasingly difficult and increasingly unrewarding.

One can see the elegance of this convergence in the quiet moment when growth gives way to maintenance. Cells re-establish stable junctions with their neighbors. They deposit extracellular matrix in a more orderly way. Blood vessels mature and become less leaky. Nerves settle into stable connections. The molecular signals that once screamed "repair" become a murmur. In the language of control theory, the system returns to a stable attractor. In the language of everyday life, the tissue feels normal again.

A regenerative animal is therefore not simply good at growing. It is good at *stopping* growth, and it stops by translating geometry, chemistry, and time into a decision that every cell can understand. A limb reaches the right length not because there is a tape measure hidden in the genome, but because cells can sense when the conditions that justify division have faded—when crowding has returned, when tension has normalized, when growth-promoting signals have been balanced by their own inhibitors, when the internal circuitry that once amplified repair cues now dampens them. The miracle, if there is one, lies in the restraint.

Chapter 4

The Electric Connection and the Mending Heart

Why does a severed salamander limb regrow while a denervated one withers? We investigate the critical 'electric' conversation between the nervous system and the blastema before shifting focus to the zebrafish, where transparency reveals how a beating heart can reverse its own biological clock to avoid scarring.

4.1 The Silent Wire

A salamander can lose a leg and grow it back with a confidence that borders on the insulting. The wound closes. A smooth cap of skin forms. Cells beneath it loosen their loyalties, stop behaving like muscle or cartilage or connective tissue, and join a swelling, pale mound of dividing cells called a *blastema*. Then the blastema does something that still feels like a conjuring trick: it builds a limb in the right order, with the right joints, in the right proportions, and it stops when the job is done.

There is, however, a quiet condition on this performance. Cut the nerves that run into the limb stump, and regeneration stalls. Not slowly, not subtly: it simply fails to get going in any meaningful way. The salamander is not confused about which shape to make. It is not making an arm instead of a leg. It is making nothing at all.

This dependence on nerves has been recognized for so long that it can feel like folklore: "the nerves are needed." But folklore can be dangerously imprecise. Nerves do two very different kinds of work. They carry electrical impulses, the rapid spikes of voltage that let animals move, sense, and react. They also carry a chemical economy: proteins released by axons and their surrounding support cells, molecules that feed, instruct, restrain, and sometimes transform the behavior of nearby tissues. If regeneration depends on nerves, which of these jobs matters? Is the limb listening to electricity, like a machine waiting for a signal on a wire? Or is the nerve more like a supply line, delivering the molecular goods that make growth possible?

A key step toward an answer came from a style of experiment that now feels almost brutally direct. In the middle decades of the twentieth century, Marcus Singer and others performed careful denervation studies in newts and salamanders. The logic was simple: amputate a limb to trigger regeneration, then surgically cut the peripheral nerves entering the stump so that the wound region is alive but no longer innervated. The result was not ambiguous. A wound epidermis could still form, but the blastema either did not develop properly or failed to progress, and the stump tended toward scarring rather than rebuilding. Put the nerves back (or allow reinnervation) and regeneration could resume.

Singer's most important contribution was not merely demonstrating dependence, but sharpening the question. If the nerve were an architect dictating form, then a denervated stump should not only fail to grow; it should also lose its sense of pattern. Yet pattern and growth proved separable in surprising ways. When denervation was performed at particular stages, some aspects of organization could still occur, but the engine of expansion sputtered. The nerve seemed less like a draftsman and more like a power source. Even that metaphor, however, risks smuggling in the wrong kind of power. A literal electrical current is not the same as biological permission.

One reason this matters is that "electric" has a seductive clarity. A wire carries current. A current does work. A limb is rebuilt: therefore, perhaps electricity rebuilds it. But living tissue is rarely that direct. Neurons are excitable, yes, and the nervous system is an electrical network. Yet the electrical spikes in nerves are brief and information-rich; they are not a steady growth stimulus. Regeneration, by contrast, is slow and materially demanding. It involves days to weeks of cell division, migration, remodeling of extracellular matrix, and the orchestration of multiple cell types. Electricity can influence cells, but it does not supply amino acids, mitotic signals, or the decision to keep dividing instead of settling into scar.

The more satisfying picture is that nerves act as part of a *niche*: the local environment that keeps regenerative cells in the right state. In a regenerating limb stump, the blastema is not an autonomous organ. It is a population of cells held in a precarious balance between possibilities. Divide too little, and the structure never forms. Divide too much, and the system risks something like a tumor. Differentiate too early, and you freeze the limb in an unfinished draft. Differentiate too late, and you may never regain proper tissue architecture. The blastema needs continuous reassurance. Nerves, it turns out, are one of the most reliable sources of that reassurance.

A nerve is more than an axon. Peripheral nerves arrive with Schwann cells that ensheath axons, with blood vessels threaded alongside, and with an ability to release growth factors both in normal life and after injury. When a limb is amputated, the severed nerves do not simply withdraw in defeat. They sprout. They regenerate toward the wound. They enter the specialized wound epidermis at the tip, a structure often called the apical epithelial cap (AEC), which is not a passive covering but an active signaling center. In many animals, epithelial tissues are powerful organizers of growth. In salamander limb regeneration, the AEC behaves like a conversation partner with the mesenchyme beneath it. The nerve helps keep that conversation going.

This "nerve dependence" is therefore not merely a dependence of limb tissue on a neural command. It is a dependence of a regenerative *system* on a missing partner. Denervation breaks multiple feedback loops at once: signals from axons to epidermis, signals from epidermis to blastema, signals from blastema back to nerve and surrounding tissues. The stump does not fail because it lacks a spark; it fails because it loses a sustaining ecology.

Modern molecular work has given this old observation a sharper edge by identifying specific factors that can stand in for nerves, at least partially, like a key that opens a door when the usual hand is absent. One of the best-known examples is a protein called nAG (newt Anterior Gradient). In the mid-2000s, work by Kumar and Gates (2007) linked denervation to loss of nAG expression in crucial tissues, including the regenerating nerve itself and the wound epidermis. nAG can act as a growth factor for blastemal cells, helping maintain their proliferative state. Even more striking, providing nAG locally could rescue aspects of regeneration in a denervated limb, allowing distal structures to be formed. This did not make the nerve irrelevant, but it made the nerve's role legible: at least part of what the nerve supplies can be replaced by a defined molecular signal.

Another example is Neuregulin-1 (NRG1), a factor well known in other contexts for its roles in development and tissue interactions. Work by Farkas and Monaghan (2016) showed that NRG1 signaling is essential for nerve-dependent axolotl limb regeneration: denervation prevents blastema formation and is associated with the loss of NRG1 and activation of its receptor pathway. Delivering NRG1 on beads could rescue regeneration in denervated limbs, in some cases all the way to digits. Again, a nerve is not a single molecule; but the ability of one factor to restart a stalled process tells you that the nerve is not simply an electrical conduit.

These findings also help correct a common misunderstanding: that nerve dependence means nerves provide *positional information*, the coordinates that tell cells whether they are "shoulder" or "wrist," "front" or "back." Positional identity is real and deeply important, but classic denervation results suggest nerves do not specify the blueprint so much as allow the blueprint to be executed. Patterning signals can persist in tissues, and blastemas can retain information about proximodistal identity even when growth is reduced. The nervous system's more consistent role is to maintain the permissive conditions for a blastema to exist and expand.

The wound epidermis is central here because it behaves like a gatekeeper. After amputation, epidermal cells rapidly migrate to cover the wound surface. In salamanders, this epidermis thickens into the AEC, which produces signals that encourage underlying cells to proliferate and remain plastic. Denervation experiments indicate that an AEC can begin to form even without axons, but it is not stably maintained. Without nerve input, the epidermal cap can fail to mature into a fully functional signaling center, and the blastema beneath it shrinks or becomes fibrotic. Put bluntly: the limb stump knows how to close a wound; it does not know how to keep a wound in the special, temporary state that regeneration requires.

That "special state" is easy to underestimate because it looks, at first glance, like the opposite of healing. In a typical mammal, the body is eager to seal, stabilize, and reinforce. Collagen is laid down. Fibroblasts invade. The immune system clears debris and pushes the tissue toward closure. These steps are sensible when the aim is survival and speed. Regeneration asks for something riskier: delay the final closure, allow a mass of proliferative cells to accumulate, and keep the environment from tipping into scar. Nerves help hold the system in that regenerative limbo.

At the cellular level, this limbo is maintained by growth signals and by restraint of differentiation. A blastema is not simply a pile of stem cells; it is a mosaic of lineage-restricted progenitors and dedifferentiated cells, depending on species and tissue, all coordinated by signals from the epidermis, nerves, and other local tissues. "Neurotrophic" factors (literally "nerve-feeding" factors) are often discussed as molecules that keep neurons alive, but in regeneration the direction of care is reversed: nerve-associated factors feed the regenerative process. The nerve becomes a nurse.

One of the most revealing twists is that nerve dependence itself is not always a fixed property. In work by Kumar and Godwin (2011) on aneurogenic limbs, limbs that develop without innervation can regenerate without nerves, at least initially, and then become nerve-dependent once they are innervated. This suggests that nerve dependence is not an immutable requirement written into the limb's cells. It can be a state imposed by developmental history. In their model, larval epidermis expresses high levels of AG-like proteins broadly; innervation down-regulates this secretory program, shifting the burden of growth support onto nerves. The implication is both unsettling and hopeful. Unsettling, because it suggests that losing regenerative ability could involve developmental "choices" that trade self-sufficiency for neural support. Hopeful, because it hints that mammalian failure might be less about missing genes and more about missing contexts.

Why would an animal ever accept such dependence? One answer is that nerves are a convenient way to coordinate growth with function. A limb that regrows without being reinnervated would be a numb, useless structure. Making regeneration contingent on nerve presence may ensure that new tissue forms only when the system can be wired back into sensation and movement. It is a kind of quality control: no nerve, no limb. The body, in effect, refuses to build an extension it cannot control.

Another answer is that nerves are a stable source of the right kinds of factors at the right time. After injury, regenerating axons and Schwann cells can release a cocktail of signals that promote cell proliferation, regulate inflammation, and influence blood vessel growth. It may be metabolically efficient to couple regeneration to the existing repair program of nerves, rather than inventing a separate, limb-specific supply chain.

Electricity still has a place in this story, but not in the simplistic way the word "wire" implies. Cells throughout the body maintain voltage differences across their membranes. Wounds disrupt these gradients and create endogenous electric fields, generated by ion transport across epithelia and by currents that flow around injury sites. Work summarized by McCaig and colleagues (2005) and later reviews has shown that applied electric fields can guide cell migration and affect wound healing outcomes, and that endogenous fields correlate with repair processes. These effects can influence the behaviors that matter for regeneration: oriented migration, cell polarity, and even the direction of neurite outgrowth. But these bioelectric cues are more like a landscape than a set of instructions. They bias cells, nudge them into alignment, and help coordinate movement. They do not, by themselves, explain why denervation collapses the blastema.

A useful way to reconcile these layers is to treat the nervous system as a mediator between electrical and molecular worlds. Nerves are electrically excitable, but their long-term influence on tissues is heavily chemical. Meanwhile, many "chemical" signals ultimately control ion channels, membrane potentials, and the electrical state of cells. Growth factors can change the expression of channels; membrane voltage can alter gene expression. The wire is silent not because there is no electricity, but because the decisive message is often carried in slow, secreted molecules rather than in rapid spikes.

One can see this silence in the timescales. A nerve impulse lasts milliseconds. A blastema persists for weeks. If nerves are required throughout, they must be doing something that remains present: maintaining the AEC, sustaining cell-cycle entry, preventing premature differentiation, supporting vascular growth, and shaping the immune response into something that clears debris without locking the tissue into scar. A single impulse cannot do that. A sustained molecular niche can.

There is also a deeper reason nerves matter, and it is emotional as much as biological: nerves are how a limb belongs to an animal. They are not an accessory. They are the limb's claim to be felt and controlled, to be part of the body's map of itself. When a salamander regrows a limb, it is not producing a detachable spare part. It is rebuilding an integrated organ that must ultimately take its place in a living circuit. Nerve dependence enforces that integration from the start.

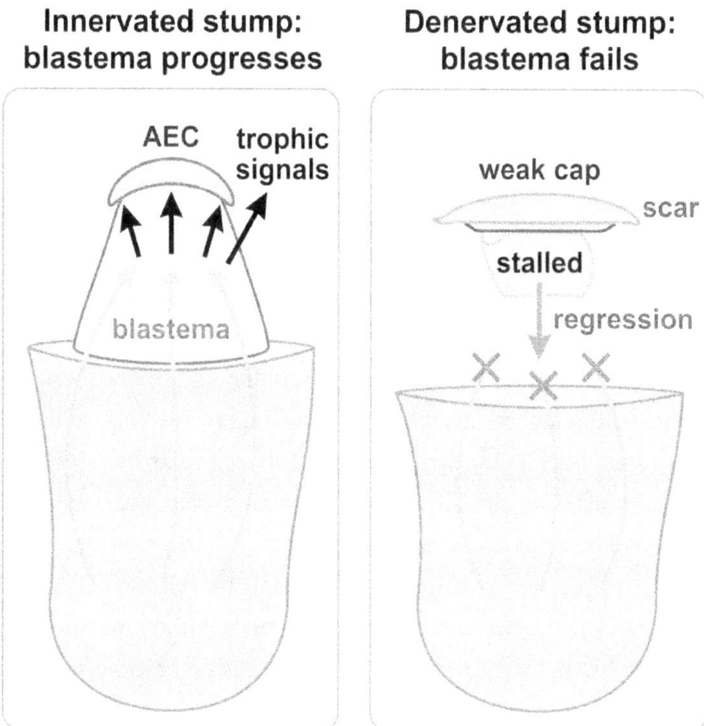

If the nerve is a supply line, the obvious next question is what exactly is being supplied. Part of the answer is specific growth factors like nAG and NRG1. Part of it is a broader support for the local environment: maintaining an epidermal signaling center, influencing immune cells, and encouraging the formation of blood vessels that will feed the growing tissue. Another part may be timing. Regeneration is not just a list of ingredients; it is a choreography. A nerve arriving early may stabilize the wound epidermis before fibrotic programs gain momentum. A nerve arriving late may find that the window has narrowed, the tissue already committed to scar.

This frames nerve dependence as a kind of biological checkpoint. The stump tests its surroundings. Are the conditions right for a risky, expansive rebuilding program? Is there an intact route for reconnection? Are the signals present that keep cells cycling but not sliding into malignancy? When the answer is yes, the blastema behaves like a development program temporarily reopened. When the answer is no, the body chooses the conservative option: close, stabilize, and move on.

There is a human echo here, not because humans regenerate limbs, but because everyone has experienced the difference between a wound that heals cleanly and one that lingers, stiffens, and leaves a thickened trace. Our own nerves influence repair in subtler ways, through sensation, blood flow regulation, and local release of factors. In some chronic wounds, especially in people with neuropathy, healing fails in part because nerves are compromised. The salamander takes that dependence and turns it into a rule: without nerve support, the wound will not be allowed to pretend it is embryonic again.

The nerve, then, is not shouting instructions. It is standing close, whispering permissions into a crowded cellular room: keep dividing, keep listening, keep the surface cap active, hold off the scar, build something worth wiring back into the self. The wire is "silent" because the sound we expect from nerves is electricity, and the sound that matters for regeneration is chemistry sustained over time, the steady presence that keeps possibility from collapsing into closure.

4.2 The Unscarred Heart

A heart attack is often described as a sudden event, but the real violence is measured in what comes after. A portion of heart muscle dies when its blood supply is cut off. In the days that follow, the body behaves like a careful builder arriving at a collapsed wall: it clears debris, lays down a temporary scaffold, and then replaces the missing architecture with something sturdy enough not to rupture. In humans, that "something" is mostly collagen-rich scar. Scar tissue is strong, but it is not muscle. It does not contract, and it does not conduct electrical impulses the way heart muscle does. The heart can compensate for a while, but scar changes the mechanics of pumping and the patterns of electrical conduction; over years, this can slide into heart failure, arrhythmias, and a narrowing of the body's tolerance for ordinary life.

The great provocation of regeneration biology is that scar is not the only possible ending. A small fish, striped and unassuming, can lose part of its heart and replace it with working muscle.

The zebrafish has become a quiet celebrity in laboratories for many reasons: it breeds readily, its embryos are transparent, and its genetics are unusually tractable for a vertebrate. Yet its most unsettling talent is not convenience; it is competence. In 2002, Kenneth Poss and colleagues reported that if you surgically remove roughly the tip of the zebrafish ventricle—about a fifth of the chamber—the fish can regrow the missing myocardium over the following weeks. The resection model was deliberately stark. This was not a small scrape. This was taking scissors to an organ that, in mammals, fails catastrophically when injured. The zebrafish survived, sealed the wound, and rebuilt.

The word *regrow* can sound like a child's fantasy. It deserves to be translated into specifics. When the ventricle is resected, blood spills and clots. A wound surface is exposed to the pericardial space. In a mammal, the next chapters are familiar: inflammation, fibroblast activation, collagen deposition, and a stiff scar that persists. In zebrafish, those early steps occur too—the fish does not skip basic triage—but they remain temporary. Collagen is laid down and then remodeled. The clot resolves. Most importantly, the missing wall is replaced by new cardiac muscle cells, called *cardiomyocytes*, that integrate with the existing tissue.

To appreciate the difference, it helps to think of the heart as two intertwined problems: mechanics and electricity. A heart is a pump made of muscle, but it is also an electrical organ. Every beat depends on an orderly wave of depolarization that travels through cardiomyocytes via specialized junctions, coordinating contraction across the chamber. Scar tissue can patch a mechanical hole, but it behaves electrically like an insulator and mechanically like a stiff patch in a flexible fabric. The result is not only weaker pumping but altered stress distribution and dangerous electrical re-entry circuits. A regenerated heart must solve both problems: replace missing muscle and restore conduction.

Zebrafish do both, and that is what makes them such a threatening counterexample to human biology. They show that adult vertebrate hearts are not universally condemned to scar.

A first guess might be that zebrafish possess a reservoir of cardiac stem cells: a hidden supply of undifferentiated cells waiting to replace damaged tissue. This idea has been attractive in many organs, including our own, because it offers a tidy narrative: damage occurs, stem cells activate, new tissue appears. But the zebrafish heart tells a more complicated, and in some ways more hopeful, story. Rather than relying primarily on an external pool of stem cells, the zebrafish seems to coax its existing cardiomyocytes back into a proliferative state.

That coaxing requires a reversal of maturity. Adult cardiomyocytes are highly specialized. They are packed with orderly contractile units called *sarcomeres*, arranged like microscopic engines. In mammals, adult cardiomyocytes rarely divide. They can grow larger; they can change gene expression; they can endure remarkable stress; but they are not eager to re-enter the cell cycle. Zebrafish cardiomyocytes, by contrast, can do something that sounds like a violation of identity: they can partially *dedifferentiate*. Dedifferentiation means that a specialized cell loosens some of the features that define its specialization, reverting toward a more youthful, flexible state. In the zebrafish heart, dedifferentiating cardiomyocytes near the injury edge disassemble portions of their sarcomeres, alter their cell–cell contacts, and re-express gene programs associated with development. This is not a full rewind to an embryonic stem cell; it is more like a skilled worker taking off heavy protective gear to climb through a narrow gap and rebuild a wall from the inside.

The evidence for this came from lineage tracing experiments that mark existing cardiomyocytes and ask a simple question: when new heart muscle appears, does it come from those marked cardiomyocytes or from some unmarked stem-like source? In 2010, Jopling and colleagues used genetic lineage tracing to show that regenerated myocardium in zebrafish largely derives from pre-existing cardiomyocytes, not from a major influx of undifferentiated progenitors. That finding did not close every debate—biology rarely grants that mercy—but it gave the field a firm center of gravity: the adult heart can be rebuilt from its own mature cells, if those cells can be persuaded to divide.

Persuasion, in biology, is signal.

One of the most revealing signals in zebrafish heart regeneration comes from a gene that developmental biologists recognize like an old friend: *gata4*. During embryonic development, *gata4* is deeply involved in heart formation. In adult zebrafish, *gata4* is largely quiet. After injury, it reawakens. Kikuchi and colleagues (2010) showed that injury activates a *gata4*-positive cardiomyocyte program, with *gata4* expression appearing in cardiomyocytes that contribute prominently to the regenerated muscle. The heart, faced with damage, reuses developmental language rather than inventing a new one.

This reuse comes with an important limitation. Development is permissive; it tolerates extensive cell division and tissue rearrangement because the embryo has little to lose. Adult tissues have more to lose: function, stability, and protection against cancer. The zebrafish heart performs a delicate compromise. Cardiomyocytes dedifferentiate enough to proliferate and migrate, but they do not become unrecognizably plastic. They remain cardiomyocytes in identity, even as they temporarily relax the rigid architecture of adulthood. The cell cycle is re-entered under tight control, and then, after the gap is filled, the cells redifferentiate, rebuilding sarcomeres and restoring the electrical and mechanical properties that make heart muscle useful.

That control is not automatic. In Poss's 2002 study, mutants deficient in a mitotic checkpoint kinase called Mps1 failed to regenerate properly and instead formed scars. The mitotic checkpoint is part of the cell's surveillance machinery: it ensures chromosomes are correctly handled during division. If you force cells to divide without such safeguards, you invite genomic chaos. The zebrafish heart's success depends, in part, on being able to run a proliferation program without losing order. That is already a clue to why mammals may be stingier with regeneration: turning on cell division in a tissue that must beat reliably every second is a frightening proposition.

Even in zebrafish, regeneration is not simply cardiomyocytes multiplying in isolation. The heart is built from layers and lineages that coordinate like a well-rehearsed ensemble. On the outside sits the *epicardium*, a thin epithelial sheet that in adult mammals is mostly quiescent. In zebrafish, the epicardium responds to injury with striking energy: it activates developmental markers, expands to cover the wound, and sends cells inward. In 2006, Lepilina and colleagues described this dynamic epicardial response and showed that a subset of epicardial cells undergoes an epithelial-to-mesenchymal transition

(EMT), a process where stationary epithelial cells become migratory and invasive. These epicardial-derived cells contribute to the formation of new vasculature, among other supportive roles. Without blood vessels, a thickening wall of new muscle would be a doomed project, starving at the center.

The importance of this support is underscored by the role of FGF (fibroblast growth factor) signaling. Blocking FGF signaling can arrest epicardial EMT and neovascularization, halting regeneration. That finding is conceptually delicious: a factor called "fibroblast growth factor" is required for a process that avoids permanent fibrosis. Names in biology are historical accidents; what matters is function. FGF here is not a scar signal but a coordination signal, linking outer-layer activation to inner-layer rebuilding.

The *endocardium*, which lines the inside of the heart chambers, also participates. It is easy to treat the heart as a lump of muscle, but muscle is only part of the organ's intelligence. Endocardial cells sense changes in flow and pressure. They respond to injury rapidly. One striking early response is the induction of retinoic acid (RA) synthesis. Retinoic acid is a derivative of vitamin A that acts as a powerful signaling molecule during development, guiding patterning and differentiation. After zebrafish heart injury, RA synthesis in endocardial and epicardial tissues is rapidly induced—within hours, through enzymes such as Raldh2—and RA signaling is required for regenerative cardiomyocyte proliferation. This is another example of developmental language being redeployed in the adult, but with careful localization and timing.

All of this creates a picture in which regeneration is less a miracle than a managed emergency. The heart responds to injury by assembling a temporary, pro-growth environment: inflammatory signals are shaped rather than allowed to harden into chronic fibrosis; extracellular matrix is deposited and then remodeled; the epicardium becomes active and contributes to revascularization; and cardiomyocytes near the wound edge are nudged into a transient, proliferative state.

Against that, set the human heart, which is not passive but stubbornly conservative. After myocardial infarction, cardiomyocyte death triggers inflammation and recruitment of fibroblasts, which deposit collagen. This process prevents rupture, which is a real risk: the heart wall is under pressure, and a weakened region can tear. A scar is a mechanical solution to a mechanical emergency. It is also a cellular solution to a biological emergency: uncontrolled proliferation in the heart would threaten not only cancer risk but immediate rhythm stability. A ventricle filled with dividing, poorly coupled cells could become an arrhythmogenic chaos. The mammalian heart's default choice is safety: patch, stiffen, survive.

That choice carries a long-term cost. Scar tissue distorts the geometry of the ventricle, increasing wall stress in adjacent muscle and promoting remodeling. The electrical discontinuities can produce dangerous rhythms. The reduced contractile mass lowers cardiac output. A person may not feel this immediately, especially if the infarct is small, but the decline can be relentless. Heart failure is not one disease but many converging consequences of a heart that has lost functional tissue and replaced it with something inert.

The zebrafish does not abolish the initial emergency response; it retools it into something temporary. That temporariness may be the entire trick. In many regenerative settings, scarring is not absent at the beginning; it is *resolved.* In zebrafish cryoinjury models, where the injury produces a more necrotic, collagen-rich lesion resembling aspects of mammalian infarction, a scar-like matrix can form and later be remodeled away as muscle regrows. The difference is not that zebrafish do not make collagen; it is that they do not accept collagen as an endpoint.

One can ask whether a fish heart is simply easier to regenerate because it is smaller, or because the fish lives in water, or because its metabolism differs. These factors may matter. But they cannot be the full explanation, because the core challenges are still present: the heart must keep beating during repair, the wound must be sealed, new tissue must be vascularized, and the electrical network must be restored.

Electrical restoration is an underappreciated benchmark. Rebuilding muscle that looks right under a microscope is not the same as rebuilding a heart that behaves like a heart. One of the most satisfying details in the zebrafish literature is the attention to functional integration. Kikuchi and colleagues (2010) reported that electrical conduction between regenerated and pre-existing myocardium is re-established on the order of weeks after injury, around two to four weeks in their resection context. That timescale matters because it hints at an intermediate state: a period when new cardiomyocytes exist but are not yet fully coupled, a potentially vulnerable phase for arrhythmias. Zebrafish traverse that phase without obvious catastrophe, suggesting that their regeneration program includes not only making new cells but making them compatible.

Compatibility involves structures called *gap junctions*, protein channels that allow ions and small molecules to pass directly between adjacent cells. In the heart, gap junctions are essential for synchronized electrical propagation. A regenerating myocardium must re-establish these connections and align the contractile apparatus so that force is transmitted efficiently. In mammals, attempts to stimulate cardiomyocyte proliferation therapeutically run into this same problem: even if you could get adult human cardiomyocytes to divide, you would still have to ensure that the new cells integrate electrically and mechanically, without increasing arrhythmia risk. The zebrafish does not magically escape physics; it negotiates it.

A historical anecdote helps here, not from fish but from the history of ideas. For decades, the dominant belief in cardiology and physiology was that adult heart muscle is "terminally differentiated," a phrase that sounds like a verdict. The heart, in this view, was a magnificent machine made of cells that were allowed to grow larger but not to multiply. Injury therefore meant permanent loss. That belief was not irrational; it was supported by the rarity of cardiomyocyte division in adult mammals and by the deadly consequences of heart failure. Yet it hardened into doctrine. Discoveries like zebrafish heart regeneration did not merely add a new animal model; they punctured a psychological barrier. Once you have seen a vertebrate heart rebuild itself, "cannot" becomes "does not," and "does not" invites "why."

The most honest answer to that "why" is not a single gene or pathway, but a network of trade-offs. Mammals have long lifespans compared to many fish, and a long-lived animal must suppress cancer over decades. It must maintain stable organ function under stress. It must avoid repair programs that risk runaway proliferation or destabilizing rhythm. The mammalian heart may have evolved to prioritize immediate structural integrity and electrical stability over the possibility of perfect restoration. Zebrafish, with different life history pressures and different physiological margins, can afford a different bargain.

That bargain is not cost-free even for the fish. Regeneration is energetically expensive. It requires coordinated signaling, immune modulation, angiogenesis, and controlled proliferation. It also demands that mature cardiomyocytes temporarily step away from their fully contractile state. During dedifferentiation, sarcomeric structure loosens; a cell that is rebuilding itself is not maximally efficient at beating. The fish accepts this temporary inefficiency for the sake of long-term restoration. Humans, faced with the risk of immediate pump failure and lethal arrhythmia, may accept permanent inefficiency for the sake of short-term survival.

One can feel the human weight of this difference by imagining two futures after a heart injury. In one, the heart is patched and the person becomes a careful custodian of a damaged organ: medications, reduced exertion, worry about rhythm, fear of the next insult. In the other, the injury is not simply survived but corrected: muscle returns, conduction normalizes, function rebounds. Zebrafish do not offer that second future to humans yet, but they prove it is biologically coherent.

The challenge is not to turn a human heart into a fish heart. It is to understand the principles that let a heart be both safe and plastic: to reawaken proliferation without unmooring identity, to recruit supportive tissues without locking into fibrosis, to lay down matrix without accepting it as permanent, and to restore electrical continuity as carefully as one restores muscle mass. The zebrafish heart does not deny the need for repair; it refuses to confuse repair with scar, and it treats function as the final test.

A scar is a kind of memory: evidence that the body chose closure over reconstruction. The zebrafish chooses memory of a different kind, one written in cells that divide, reconnect, and resume the work of beating in synchrony, as if the injury were a temporary interruption rather than a lifelong compromise.

4.3 Windows into the Machine

There is a particular frustration that comes with studying regeneration in most animals: the action happens behind closed doors. A salamander's limb stump swells, the skin smooths, and days later a bud appears, but the decisive moments—which cells decided to divide, which stayed put, which changed identity, which called in blood vessels, which pushed the tissue toward scar—are hidden in depth. A mouse heart remodels after injury, yet the critical argument between repair and regeneration takes place inside an opaque, beating organ. For a long time, the scientist's view of these events resembled forensic work. You sacrificed an animal at a chosen time point, sliced tissues into thin sections, stained them, and inferred a story from snapshots. It was honest work, but it encouraged a particular kind of illusion: that regeneration proceeds in clean stages, one after another, as if every cell had read the same timetable.

Zebrafish do something rare for a vertebrate: they offer a window. At the earliest stages of life, zebrafish embryos and larvae are naturally transparent. Their organs are not only visible; they are accessible to light in a way that makes cellular behavior recordable. You can watch a heart form and beat in a living animal that is only a few millimeters long. You can label a subset of cells and follow their descendants as they migrate, divide, and integrate. You can injure tissue and, instead of guessing who responded, see it happen.

This has changed regeneration biology in the same way that time-lapse photography changed our understanding of plant growth. Before time-lapse, it was easy to think of plants as still. After time-lapse, it became impossible not to notice their movement. Something similar happens when you watch a zebrafish heart regenerate under a microscope. Injury is no longer an abstract trigger; it is a physical event followed by a choreography of living cells, each making choices that can be measured.

The most famous early "movie" in zebrafish work is not regeneration but development: the formation of the heart tube, the looping of chambers, the onset of rhythmic contraction. This mattered because it trained a generation of biologists to think of the heart as observable. When heart regeneration research accelerated after Poss's 2002 demonstration that the zebrafish ventricle can regrow, the field already had a culture of imaging. The question became not only *does* the heart regenerate, but *how* can you see the decision points that make regeneration win over scar?

Transparency also changes what counts as evidence. In a fixed tissue section, a proliferating cardiomyocyte is a rare, precious sight. You can stain for a cell-cycle marker and count positive nuclei, but you are always haunted by alternative explanations: was this truly a cardiomyocyte, or a neighboring cell? Did it divide, or merely express a marker? Was it near the injury at the right time, or did it drift there later? Live imaging does not abolish these questions, but it lets you answer them with more than inference. You can watch a labeled cardiomyocyte enter the cell cycle, round up, divide, and then spread back out into the myocardium, all while the heart continues to beat. The event stops being a rumor in the tissue and becomes a witnessed act.

To do that, researchers needed a way to make specific cells glow. Zebrafish genetics makes this unusually straightforward. By attaching fluorescent proteins—green fluorescent protein (GFP) and its many color variants—to promoters active in particular cell types, one can label cardiomyocytes, epicardial cells, endocardial cells, immune cells, or fibroblast-like populations. These labels can be stable and heritable, allowing long-term tracking. More sophisticated tools allow lineage tracing: marking a cell at a chosen time and then identifying all of its descendants later. This is how lineage studies in the zebrafish heart could argue, with unusual clarity, that regenerated myocardium arises largely from pre-existing cardiomyocytes rather than from an uncommitted stem-cell pool.

Yet seeing is not only about labeling. It is also about getting light into tissue without damaging it. Traditional confocal microscopy can produce beautiful images, but it does so by scanning a point of light through a specimen, which can be slow and can expose tissues to enough light to cause phototoxicity. Regeneration is not a single moment; it is an extended process. If you want to watch it unfold in real time, you need imaging methods that are gentle enough for living tissues.

This is where light-sheet microscopy became a quiet revolution. In light-sheet imaging, a thin sheet of light illuminates only one plane of the specimen at a time, and the camera collects fluorescence from that plane. Because you avoid illuminating out-of-focus regions, you reduce unnecessary exposure and can image faster with less damage. In translucent zebrafish embryos and larvae, light-sheet microscopy makes it possible to record three-dimensional stacks over time—four-dimensional imaging, in the jargon—capturing not just the location of cells but their movements, shape changes, and divisions. Applied to a beating heart, this creates a further challenge: motion.

A beating heart is the enemy of sharp images. Even if the fish is tiny, the heart wall moves rapidly, and the same cell can be in a different position from one frame to the next. This could have been a fatal limitation, but it became an engineering problem instead. Researchers developed methods to synchronize image acquisition to the cardiac cycle, to computationally "gate" images, and to reconstruct a stable representation of a beating organ. The heart became an observable machine without being stopped.

This matters because regeneration is not simply the addition of new cells. It is the restoration of function, and function in the heart depends on rhythm. Being able to watch the regenerating heart beat, rather than seeing it only as fixed tissue, keeps the narrative honest. A patch of new myocardium that looks correct but contracts out of sync is not regeneration in any meaningful sense.

One way to keep function in view is to measure electrical integration, not just structure. Cardiac electrical activity can be visualized with optical tools: fluorescent reporters that respond to voltage changes or calcium transients. Calcium is particularly useful because calcium influx is intimately tied to contraction, and calcium-sensitive dyes or genetically encoded calcium indicators can reveal waves of activation sweeping through the heart. In adult zebrafish ventricles, optical voltage mapping has been used to show that conduction between existing and regenerated myocardium can be re-established within weeks after injury, on the order of two to four weeks in resection models. This kind of measurement anchors regeneration to a functional endpoint that matters: not merely the replacement of tissue, but the reconnection of a living electrical network.

There is a second kind of transparency that has expanded these windows further: adult zebrafish that remain see-through. Normally, adult fish are opaque because they develop pigment cells that scatter and absorb light. But genetic mutants exist that lack key pigment cell types. By combining such mutations, researchers created "casper" zebrafish—transparent adults named with a wink for the friendly ghost. In casper fish, deep tissues can be observed for longer periods in living adults, allowing longitudinal imaging of processes that unfold over weeks. Heart regeneration, which in zebrafish occurs on a timescale far longer than embryonic development, becomes more tractable when you can return to the same animal repeatedly, watching the same injury site change.

The emotional difference between sacrificing a series of animals and revisiting the same animal is subtle but real. In the first approach, each time point is a separate individual, and variability is an unavoidable haze. In the second, the animal becomes a continuing narrative. You can ask whether one fish's heart is regenerating faster than another's, whether a particular intervention accelerates integration, whether scarring resolves smoothly or lingers. Longitudinal observation reduces the temptation to pretend that biology is more uniform than it is.

If imaging provides the window, reporters provide the vocabulary.

One of the most useful vocabularies in regeneration is the cell cycle, because regeneration so often hinges on who divides and who does not. A heart that replaces lost muscle must coax cardiomyocytes into proliferation. A limb blastema must sustain high proliferation without losing control. Yet the cell cycle is not visible to the naked eye. It is a series of molecular states: G1 (growth and preparation), S (DNA synthesis), G2 (preparation for division), and M (mitosis). Traditionally, one inferred these states with stains on fixed tissue. Fluorescent reporters made them observable in living animals.

The FUCCI system—Fluorescent Ubiquitination-based Cell Cycle Indicator—is a particularly elegant example. It uses proteins that are naturally degraded at specific phases of the cell cycle, fused to fluorescent tags, so that a cell changes color depending on where it is in the cycle. A cell in G1 might glow red; as it enters S phase, it shifts toward green. The details can vary with the exact implementation, but the central delight remains: the cell cycle becomes a color map. In a transparent embryo, you can watch cells transition through states in real time, rather than counting static markers.

That capability does more than satisfy curiosity; it makes arguments possible. Consider the zebrafish heart again. If cardiomyocytes near the injury edge are truly re-entering the cell cycle, you should see the color transitions in those cells, localized in space and time. If a genetic mutation or a drug blocks regeneration, you can ask whether it blocks dedifferentiation, blocks cell-cycle entry, blocks progression through mitosis, or blocks later redifferentiation. Each of those failures has a different meaning and suggests a different constraint. Without live imaging, these steps blur together.

There is a historical echo here from an older era of biology, when the microscope first turned tissues into organized worlds. Early histologists were astonished to discover that organs were made of cells with distinct shapes and arrangements, rather than being homogeneous substances. Imaging in zebrafish has done something similar for regeneration: it has turned what looked like an organ-level ability into a set of cellular behaviors that can be watched and quantified. The microscope did not merely reveal structure; it revealed that structure is dynamic. A regenerating heart is not a sculpted object. It is a shifting negotiation among cell types.

Imaging has also changed how researchers think about the supporting cast. In the zebrafish heart, cardiomyocytes do not regenerate alone. Epicardial cells spread and activate. Endocardial cells respond rapidly to injury and participate in signaling. Immune cells arrive and depart, shaping inflammation. Blood vessels regrow. Extracellular matrix is laid down and remodeled. In fixed tissue, these players can be mapped, but in live imaging their timing becomes obvious. Some arrive early and set conditions; others arrive later and consolidate. A pro-regenerative immune response, for example, might be characterized not only by the presence of certain cells but by their dwell time and behavior at the wound edge. A fibrotic outcome might be associated with immune cells that linger too long or signal too strongly for collagen deposition. These are hypotheses that can be tested when one can watch the cells rather than only count them.

The same principle applies to nerves and bioelectric cues, even in contexts where nerves are not the main actor. In a limb, nerve fibers can be visualized as they sprout into a wound. In zebrafish larvae, axons and glia can be fluorescently labeled, revealing how innervation responds to injury. Electrical activity in the heart can be read out optically. The earlier idea of "the silent wire" becomes more than metaphor: one can watch the physical wire of innervation approach and one can watch the electrical waves that must ultimately be restored.

There is also a more philosophical shift that comes from these windows: regeneration becomes less like a special trick and more like a decision made repeatedly. A cardiomyocyte at the injury border does not make one grand choice to "regenerate." It makes smaller choices: loosen sarcomeres, change gene expression, re-enter the cell cycle, divide, migrate, reconnect. Each choice can be influenced by signals from neighboring cells and by the physical environment. Imaging makes these choices visible, and visibility makes them vulnerable to intervention.

This is where the story begins to touch medicine without pretending it is simple. The zebrafish heart's ability to regenerate does not immediately translate into a therapy for human heart failure. Human hearts are larger, under different pressures, and more dependent on immediate, stable conduction. Human cardiomyocytes are more reluctant to divide, and the human immune system, while powerful, is not tuned for regeneration. But zebrafish imaging has already contributed something indispensable: it has provided a way to screen, to test, and to falsify.

A drug that seemed "pro-regenerative" when judged by a final histological image might, under live imaging, be revealed as chaotic: it induces proliferation but disrupts coupling, or it accelerates cell cycle entry but increases cell death. Another intervention might appear modest in a snapshot but, in a movie, show improved coordination: faster revascularization, more orderly cardiomyocyte migration, earlier restoration of conduction. Movies are not merely prettier; they change the criteria by which success is judged.

The transparency of zebrafish embryos also makes it possible to study regeneration in tissues beyond the heart in a unified way: fin regeneration, spinal cord repair, vascular regrowth, and wound healing. This matters because regeneration is rarely organ-specific in its logic. The same families of signals—growth factors, extracellular matrix cues, immune modulators, bioelectric gradients—are reused across contexts. Watching these processes in embryos allows researchers to identify conserved behaviors and then ask whether adult organs can be coaxed into similar states. The embryo becomes a reference for what plasticity looks like when it is normal.

One can feel the power of these windows in a very concrete laboratory scene. A researcher induces a small injury in a zebrafish larva and mounts the animal in a drop of gel under a microscope. The heart continues to beat. Fluorescent cardiomyocytes outline the chamber. Over minutes and hours, cells at the injury edge change shape. Over days, a previously empty region fills with labeled descendants. The animal is not sacrificed to extract its truth; it carries its truth in motion. The researcher can return the next day and watch what has changed. That continuity is a kind of intimacy with biology that older methods could not offer.

In the end, transparency is not merely optical. It is conceptual. When you can see the process, you stop treating regeneration as a black box. You stop accepting vague verbs like "heals" and "repairs" and "regenerates" as explanations. You begin to ask which cells moved, which divided, which signals turned on, which electrical waves returned, and how the tissue avoided taking the comfortable path into scar.

A zebrafish embryo, almost invisible in a dish, can reveal a truth that is easy to miss in larger animals: regeneration is not a single capacity. It is a sequence of reversible commitments, monitored and corrected in real time. Under the microscope, the living heart is not just an organ; it is a set of negotiations that can be watched as they happen, and that simple visibility has become one of the most powerful tools for turning the mystery of regeneration into something testable.

Chapter 5

The Wall of Scars

Evolution made a calculated trade-off for mammals: survive the infection today rather than regrow the limb next month. This chapter uncovers how the very mechanisms that save us from bleeding to death-rapid fibrosis and aggressive immunity-simultaneously erect a biological barrier against complex tissue repair.

5.1 The Speed Trap

A clean cut across a finger has an uncanny power to shrink the world. The bright bead of blood is small, but it reorganizes your attention with the authority of an alarm. Within seconds, the body begins acting as if the only story that matters is containment: stop the leak, wall off the breach, keep microbes out. Long before there is any chance to rebuild what was lost, there is the more urgent problem of staying alive for the next ten minutes.

That priority is so deeply baked into mammalian biology that it can be hard to imagine an alternative. Yet it exists. A salamander can lose a limb and, rather than slapping on a quick patch, can stage a slow reconstruction that respects the original blueprint: bones reappear in the right order, muscles reattach, nerves rethread themselves. The difference is not that salamanders enjoy pain or tolerate danger better. The difference is that mammals live in a "speed trap" where time is a predator. We are warm, high-pressure vessels that run on rich blood and dense oxygen delivery. We are also home to an immense microbial world that is mostly harmless until it finds a way into places it does not belong. In that context, wound healing becomes less like careful carpentry and more like emergency engineering.

Two killers haunt open tissue: hemorrhage and infection. Blood loss can be rapid and unforgiving. The circulatory system of a mouse or a human is not a gentle stream; it is a pressurized network designed to keep oxygen arriving on schedule. A hole in a vessel is not simply a missing part to be replaced later. It is a leak in a system that depends on closed pipes. Infection is slower than bleeding, but just as decisive. Skin is not merely a wrapper; it is a boundary with the outside world, an ecological border checkpoint. Breach it, and organisms that are normally restricted to surfaces can enter deeper tissue, where they meet nutrients, warmth, and relative darkness. Evolution does not reward a perfect regrowth that arrives too late. It rewards the animal that seals the gap and walks away.

The first responders are mechanical and chemical. Damaged vessels constrict. Platelets, those tiny anucleate cell fragments that circulate like repair kits, stick to exposed collagen and to each other, forming a plug. A cascade of clotting proteins converts soluble fibrinogen into fibrin strands that crosslink into a net. The clot is a provisional dam: imperfect, fragile, and absolutely necessary. It does not restore function. It buys time.

Then the living crews arrive. Immune cells flood the area, drawn by alarm signals released from ruptured cells and by chemical gradients that spread from the wound like smoke. They clear debris and attack microbes. At the same time, the body begins laying down a temporary scaffold: a wet, protein-rich matrix in the wound bed that cells can crawl across. This is where the speed trap tightens. The body now faces a fork in the road. It could keep the matrix loose and permissive, allowing cells to reorganize, to negotiate the geometry of the missing tissue, and to re-pattern the area. Or it could stabilize the site quickly, strengthening the scaffold into something tough enough to withstand movement, tension, and bacterial assault.

Mammals typically choose toughness.

The main builders of that toughness are fibroblasts, the workhorse cells of connective tissue. Under normal conditions, fibroblasts are quiet custodians. They maintain the extracellular matrix, the meshwork of proteins and sugars that holds tissues together and conveys physical forces. After injury, fibroblasts become urgent. They migrate into the wound bed, proliferate, and begin depositing collagen in large amounts. Collagen is the body's structural polymer, a rope-like protein that can be woven into a fabric strong enough to resist tearing. In a wound, collagen behaves like emergency rebar and concrete poured into a crater: it fills space, restores tensile strength, and reduces the chances that the wound will reopen. The result, once remodeled, is scar tissue.

A scar can look deceptively simple from the outside: a pale line, a smooth patch, a slightly raised ridge. On the inside it is a radically different architecture. In uninjured skin, collagen fibers are arranged in a basket-weave pattern that allows the tissue to stretch and recoil. In many scars, fibers are thicker, more aligned, and more densely packed. This density is a mechanical triumph and a biological obstacle. It makes the repaired area strong, but it also creates a physical environment that resists the kind of cellular choreography required for regeneration.

A useful way to think about this is to separate two goals that are often conflated. *Closure* is sealing the surface and restoring a barrier. *Reconstruction* is rebuilding the original structure with the right parts in the right places. Closure can be fast and ugly. Reconstruction is slow and precise. Mammals are built to win the race to closure because losing that race can be fatal.

The speed trap has an evolutionary logic that shows up vividly in history, especially in the age before antibiotics. Consider the injuries of nineteenth-century warfare. Surgeons working through the Napoleonic campaigns, and later the American Civil War, saw countless wounds that were survivable in terms of immediate tissue loss but lethal because of infection. A musket ball could tear flesh without instantly killing; what often followed was suppuration, fever, and death days later. When Joseph Lister began promoting antiseptic surgery in the 1860s, he was responding to a brutal arithmetic: keeping microbes out of wounds transformed outcomes. In that era, an organism that sealed wounds quickly and erected an inflammatory, fibrotic barrier would have enjoyed a powerful selective advantage. Sterility was not a cultural practice; it was a biological gamble.

Modern medicine, with sutures and antibiotics and sterile operating rooms, can sometimes give tissues the kind of protected environment that nature rarely provides. Yet even in a perfectly clean surgical incision, the mammalian body still runs its ancient program. It closes, it fortifies, it scars. The fact that we can protect wounds externally only highlights how deeply internal priorities are entrenched.

One of the most vivid demonstrations that scarring is a choice rather than an inevitability comes from a comparison across mammals themselves. Most laboratory mice heal skin wounds with prominent scarring. But certain spiny mice (often discussed under the genus *Acomys*) can heal some skin injuries with remarkably little scar, even regenerating structures like hair follicles in contexts where ordinary mice do not. This is not salamander-level limb regeneration, but it is enough to make the central point uncomfortable: the mammalian toolbox contains more than one setting. The default setting emphasizes rapid stabilization, but it can be tuned.

Why does tuning matter? Because scar is not just *extra collagen*. Scar is a different *state* of the tissue, and it does not merely occupy space; it dictates what kinds of cells can enter, how they behave once they arrive, and what signals they can hear.

The wound bed is a crowded negotiation. Cells do not move through empty air. They move through the extracellular matrix, gripping it with surface receptors, pulling against it, sensing its stiffness, and responding to the chemical tags embedded within it. A loose, hydrated matrix is like a flexible climbing net. It permits migration, remodeling, and the kind of collective rearrangement that is needed if tissue is to re-pattern itself. A dense, crosslinked collagen patch is more like a cured epoxy. Cells can attach to it, but it resists being reorganized. It tells cells, through mechanical cues, to stop behaving like flexible builders and start behaving like contractors finishing a job under strict safety codes.

In mammalian skin wounds, a key step in rapid closure is contraction. Specialized fibroblasts called myofibroblasts develop contractile machinery similar to that used by muscle cells. They grab the matrix and pull, drawing wound edges together like a tightening drawstring. Contraction is efficient. It reduces the area that must be filled, speeds re-epithelialization (the crawling of skin cells across the surface), and lowers the window of vulnerability to infection. But contraction also distorts geometry. It is excellent for sealing a gap and poor for restoring the original layout. Where regeneration must preserve a map, contraction is content to crumple it so long as the border closes.

From the perspective of engineering, scar is a structural patch: a rapidly fabricated composite designed for load-bearing. From the perspective of pattern, scar is a barricade. It changes the local landscape so drastically that the cells capable of regeneration (or of adopting regenerative roles) face an environment that discourages precisely the behaviors regeneration requires.

That becomes clearer if one imagines the early wound as a kind of construction site with two possible permits. Under a "repair permit," the priority is to restore safety and structural integrity as fast as possible. Under a "regeneration permit," the priority is to restore the original design, even if it takes longer. Mammals tend to issue the repair permit almost automatically. The contractors show up, pour collagen, and harden the site. Once the concrete cures, it becomes difficult to excavate without risking collapse.

The trouble is not that collagen exists; collagen is indispensable. The trouble is timing and organization. In regeneration, a provisional matrix can be laid down and then remodeled in a way that remains permissive to pattern formation. In scarring, the matrix rapidly becomes dense and aligned, and the longer it persists, the more it recruits a whole suite of reinforcing mechanisms: additional collagen deposition, crosslinking enzymes that stiffen the fibers, and signaling loops that keep fibroblasts in an activated state. A temporary scaffold turns into a permanent wall.

This wall is not only physical. It also changes chemistry. The extracellular matrix binds growth factors and presents them to cells. Different matrix compositions can amplify or dampen the same signal. A wound bed rich in certain collagen types, proteoglycans, and glycoproteins can bias cells toward migration and proliferation or toward quiescence and contraction. When the matrix stiffens, it alters how cells organize their internal skeletons. That mechanical change feeds back into gene expression. A cell on a soft surface behaves differently from the same cell on a rigid one, not by metaphor but through measurable pathways that convert force into transcription.

All of this matters because regeneration is not merely "more growth." It is growth that respects a plan. In the regenerating limb of a salamander, cells form a structure called a blastema, a mound of proliferative cells capable of rebuilding multiple tissue types in the correct spatial order. The blastema is not magic; it is a regulated, transient state that requires access to signals and room to reorganize. A dense scar, deposited early and maintained, makes both access and room scarce. It is hard to form a blastema-like environment when the wound bed has already been reinforced against change.

Even within mammals, the contrast between fetal and adult wound healing underscores the point. Human fetuses can heal certain skin wounds with minimal scarring early in development. Later, the same kinds of injuries produce scars. The fetus is not simply "better at healing." It lives in a different ecological and mechanical context: a sterile amniotic environment with different immune dynamics, different skin structure, and different constraints. The adult lives in a world where every cut is an invitation to microbes, and where movement and mechanical stress constantly threaten to reopen the breach. The adult's healing program responds accordingly.

One can see the speed trap in the way mammalian wounds tolerate imperfection. A scar may lack sweat glands and hair follicles. It may have altered pigmentation and reduced flexibility. Yet it is closed, strong, and resistant to infection. It is a compromise that favors survival and rapid return to function. For an animal that must keep moving to find food, avoid predators, or care for offspring, a "good enough" patch is often better than a slow perfection.

The compromise becomes particularly poignant when scars occur in tissues where architecture is inseparable from function. In the heart, for example, a scar is not merely cosmetic; it can disrupt electrical conduction and reduce pumping efficiency. After a heart attack, a region of cardiac muscle dies from lack of blood supply. The body cannot afford to leave a hole in the ventricular wall, so fibroblasts deposit collagen to prevent rupture. That collagen patch can keep the heart intact while also leaving it weaker and more prone to arrhythmias and failure. The very mechanism that saves a life in the acute phase can shadow the rest of it.

There is a quiet irony here. Mammals are sometimes described as having "poor regeneration," as if it were a deficiency. But the mammalian wound program is extraordinarily competent at what it evolved to do: keep a high-pressure, microbe-exposed organism alive after injury. It is not a broken regeneration program. It is a different optimization problem, solved under severe time constraints.

A helpful comparison comes from materials science. Imagine two ways to repair a cracked ceramic bowl. One way is to apply epoxy immediately: the bowl becomes usable quickly, but the repair line remains visible and the material properties at the seam are different. Another way is to re-fire the ceramic with fresh clay integrated into the original structure: the result can be more seamless, but the process requires time, controlled conditions, and the bowl must be out of service. Mammals are built to reach for epoxy because the world does not pause while tissues are re-fired.

The speed trap also clarifies why some attempts to coax regeneration in mammals run into a stubborn barrier. If the scar program has already poured its concrete, adding growth factors later is like trying to plant a garden on asphalt. Cells might proliferate, but they cannot easily re-pattern the region, reconnect structures, and recreate the original microenvironment. The timing of events matters. A wound that closes quickly and stabilizes early is, in a sense, declaring: "No more rearrangements."

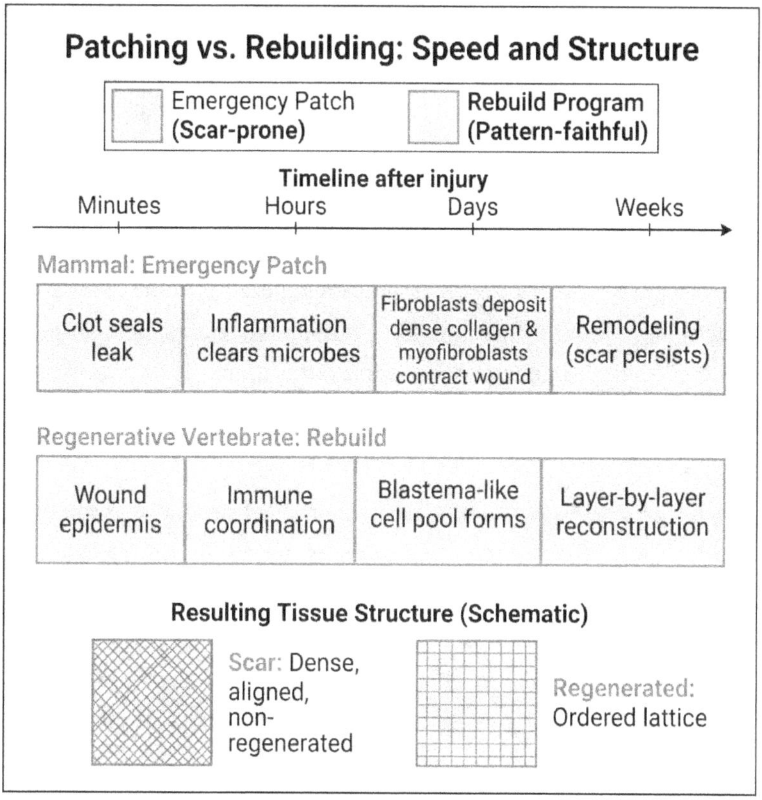

The diagram is a simplification, but it captures the crucial asymmetry: the scar-prone route front-loads mechanical stability, while the regenerative route front-loads patterning competence. The mammalian decision is not made by a single molecule flipping a switch. It emerges from many pressures acting in the same direction: rapid clotting, rapid inflammation, rapid fibroblast activation, rapid matrix deposition, rapid contraction. Each step makes the next step more likely, and together they create a self-reinforcing trajectory. If a tissue is to regenerate, it must keep open a window in which cells can reorganize without the scaffolding becoming a prison.

Everyday life offers small, personal evidence of how relentlessly mammals protect that window from staying open. Pick at a scab and you will learn how strongly the body resists delay. The wound oozes, the clot reforms, the surface reseals. The body treats interference as a renewed emergency. It is not being petty. It is responding to a world in which an unsealed wound is a dangerous interface.

This is why the question "Why can't we regenerate like salamanders?" can be misleading. It assumes regeneration is the default, and scarring is a failure. In mammals, scarring is a successful emergency strategy that hardens too soon to permit a more elegant reconstruction. That hardening is not a mere side effect. It is the point: build a wall quickly, even if it is ugly, because the alternative may be death.

Once that wall exists, living tissue can adapt around it, sometimes impressively. Skin regains sensation as nerves sprout into the area. Strength improves with remodeling. In some organs, remaining cells enlarge or increase workload to compensate. But adaptation is not the same as restoration. The scar remains as a record of the speed trap: a line drawn by evolution, marking the place where survival outranked symmetry.

5.2 The Immune Double-Agent

A scar is often blamed on collagen, as if the body simply got overenthusiastic with its glue gun. The deeper culprit is less tangible and far more human in its logic: suspicion. The immune system does not know, in the abstract, whether a wound is a clean surgical incision or a bite that dragged dirt and saliva into muscle. It treats open tissue as a crime scene with missing evidence. The safest assumption is that an invader is present, multiplying, adapting, and prepared to exploit any delay. So the immune system arrives as a demolition crew: it floods the site with weapons, corrosive chemistry, and blunt-force tactics designed to kill bacteria quickly.

That demolition is lifesaving. It is also the beginning of a trap. The same immune system is supposed to become an architect: to quiet the battlefield, coordinate rebuilding, and help the tissue return as close as possible to what it was. Regeneration, where it happens, depends on this pivot. The puzzle is not that mammals mount strong inflammation. The puzzle is that mammals so often fail to stop, or fail to switch cleanly, leaving the wound in a prolonged state of hostility where rebuilding becomes impossible or deeply distorted.

One can think of wound healing as a conversation with two volumes: the loud, urgent language of danger and the quieter, patient language of repair. In mammals, the danger channel can drown out everything else. That is the immune double-agent: the same system that protects an injured body can also sabotage its chance to rebuild.

The cast of immune cells arriving at a wound is large, but a single cell type has become central to the modern story of regeneration: the macrophage. Macrophages are professional eaters. Their name comes from Greek roots meaning "big devourer," and it is not an exaggeration. They engulf bacteria, dead cells, and fragments of wrecked tissue. They also release signals that recruit other immune cells, instruct fibroblasts, influence blood vessel growth, and coax stem or progenitor cells into action. Macrophages are not merely janitors; they are managers.

In the lab, immunologists often describe macrophages in terms of a simplified spectrum. At one end are macrophages in an inflammatory mode, sometimes called M1-like: aggressive, microbe-killing, and prone to releasing inflammatory cytokines (chemical messages between cells) and reactive molecules. At the other end are macrophages in a repair mode, sometimes called M2-like: more focused on resolving inflammation, promoting tissue remodeling, and supporting new blood vessels. Real wounds contain macrophages in many intermediate states, and the labels can mislead if taken too literally. Yet the underlying idea is robust and important: macrophages must change their behavior over time for healing to proceed well.

A clean switch is hard, because the wound itself keeps changing. Bacteria may be present early and gone later. Dead tissue must be cleared. Oxygen levels fluctuate. Mechanical forces tug on the site. The immune system is trying to infer what stage it is in, and it must do so without the luxury of certainty. The cost of switching too early can be catastrophic: a surviving bacterial colony can become an abscess, a systemic infection, a death sentence in a pre-antibiotic world. The cost of switching too late is also substantial: prolonged inflammation encourages fibrosis, the overgrowth of connective tissue that produces thick scars and, in organs, can cripple function.

Historically, doctors learned the stakes of that timing long before they understood macrophages. In the late nineteenth century, Elie Metchnikoff watched motile cells in starfish larvae and other organisms engulf foreign particles, and he argued that these cells were central to immunity. His idea, phagocytosis, helped lay the foundations of cellular immunology. Metchnikoff's phagocytes were celebrated as defenders. What was less appreciated at the time was that these defenders would later be recognized as conductors of repair as well. The same cell that eats bacteria can decide whether a wound becomes a thin line or a stiff tether.

The clearest evidence that macrophages are not optional extras in regeneration comes from salamanders. In the early 2010s, James Godwin and colleagues performed an experiment in axolotls that reads like a detective story with an unexpectedly simple culprit. They depleted macrophages around the time of limb amputation. The animals still closed the wound. They did not bleed to death. The surface sealed. Yet regeneration failed, and instead of a new limb, the axolotl formed a fibrotic scar-like tissue. Closure without regeneration is a brutal separation of two processes that are usually blurred together.

It demonstrates that macrophages, present at the right time, are part of what prevents the wound from hardening into a barrier.

A similar logic appears in the salamander heart. Godwin's group (2017) showed that macrophage depletion early after heart injury did not merely slow repair; it steered the tissue into a different destiny, producing a permanent, highly cross-linked extracellular matrix scar. Even with other pro-growth events continuing, the scar consolidated. The result suggests that regeneration is not just about making muscle cells divide. It is also about preventing the environment from becoming unreceptive, an environment that immune cells help shape.

These salamander results are unsettling because they invert the usual mammalian narrative. In many mammalian discussions, inflammation is treated primarily as damage: something that destroys tissue and must be suppressed. In salamanders, the right immune activity is permissive, even necessary. The difference is not that salamanders lack inflammation. The difference is that their immune response seems better at changing gears and, crucially, at preventing fibroblasts from locking the wound into a fibrotic state.

It is tempting to romanticize the salamander immune system as gentler or more enlightened. It is neither. It is simply tuned to a different ecological compromise. A salamander can afford a slower reconstruction and can keep the wound in a more plastic state for longer, while still surviving. Mammals, as high-pressure, microbe-exposed organisms, often respond as if every wound is an emergency that could become septic. When the immune system stays in demolition mode, it is not being foolish; it is playing the odds.

Those odds change dramatically with chronic inflammation, one of the defining medical themes of modern life. A wound is meant to be a temporary crisis. But in many conditions, the body behaves as if the crisis never fully ends. Diabetes is a familiar example: high glucose can impair blood flow, damage small vessels, alter immune cell behavior, and slow healing. Chronic wounds in diabetes are notoriously difficult to treat and can lead to infection and amputation. The immune system is active, but it is active in a way that does not produce resolution. The wound becomes a place where inflammation and tissue breakdown coexist with frustrated attempts at repair. Under those conditions, regeneration is not merely unlikely; it is actively forbidden by the chemistry of the site.

The language of demolition is chemical as much as cellular. Inflammatory immune cells release reactive oxygen species and enzymes that can kill microbes but also degrade tissue. They secrete cytokines like tumor necrosis factor and interleukins that amplify the immune response and recruit more cells. These signals are powerful precisely because they can mobilize a whole system quickly. But powerful signals are dangerous when they persist. They keep fibroblasts activated. They encourage myofibroblasts to remain at the wound, pulling and laying down collagen. They influence blood vessel growth, sometimes leading to leaky, disorganized vasculature that sustains inflammation. They can even alter nerve regrowth. A wound under persistent inflammatory signaling becomes a neighborhood where no one invests in long-term planning.

The immune-to-fibroblast connection is especially important. Fibroblasts are often portrayed as passive responders, but they are deeply interactive. They can sense inflammatory signals and respond by producing extracellular matrix. They can also send their own signals back, shaping immune behavior. In fibrosis, this cross-talk can become a vicious circle: inflammation activates fibroblasts; activated fibroblasts change the matrix and secrete factors that keep immune cells engaged; the stiffened matrix mechanically reinforces fibroblast activation. This is how a temporary patch becomes an entrenched wall.

The immune double-agent is therefore not a matter of "good cells" and "bad cells." It is a matter of timing, context, and feedback loops. Early inflammation is a necessary firebreak. Later, persistent inflammation is a fire that keeps smoldering in the walls.

The idea that immunity can be tuned toward regeneration is not confined to amphibians. Mammalian exceptions, and mammalian experiments, hint that the program is pliable. Spiny mice, with their unusually regenerative skin responses, show a different balance of inflammatory signals after wounding compared with ordinary laboratory mice. The wound environment appears less dominated by the classic inflammatory surge and more compatible with remodeling that resembles developmental or fetal-like patterns. The point is not that spiny mice have no immune response; it is that their immune response seems less likely to tip into a self-sustaining fibrotic loop.

More direct evidence comes from targeted manipulations in laboratory mice. In 2024, Huang and colleagues reported that knocking out a chemokine receptor called CXCR2 could promote scarless regeneration across several injury models, and that circulating factors, including elevated granulocyte colony-stimulating factor (G-CSF), could transfer aspects of this regenerative healing tendency. G-CSF is best known clinically for boosting certain white blood cells, and the idea that it could tilt wound outcomes is provocative. The results suggest that systemic immune tone is not just background noise; it can set the healing trajectory. In other words, the immune system does not merely respond to wounds. It brings a pre-existing mindset to them.

Anti-inflammatory cytokines also reveal leverage points. Interleukin-10 (IL-10) is a molecule with an almost diplomatic reputation: it dampens inflammatory responses and encourages resolution. In 2008, Peranteau and colleagues showed that increasing IL-10 in an adult wound model could reduce inflammatory mediators and shift healing toward a more regenerative phenotype. This resonates with the broader observation that fetal wound healing, which can be relatively scarless, occurs in an environment with a different inflammatory balance. The implication is not that "less inflammation" is always better. It is that the *shape* of inflammation over time matters, and that guiding the descent from demolition to reconstruction is one of the keys to avoiding the scar wall.

At a dinner party, someone might ask the obvious question: why not simply suppress inflammation in wounds and let regeneration happen? The reason is that inflammation is not an accessory; it is a gatekeeper for survival. Suppress it indiscriminately and infections rise. The immune system's demolition tools are not optional in a world full of microbes. The more realistic goal is not to silence the immune system, but to help it complete its natural arc: attack early, then stand down, then coordinate rebuilding.

Standing down is not passivity. It is a different kind of work. Resolving inflammation involves clearing spent immune cells, cleaning up molecular debris, stabilizing blood vessels, and promoting the orderly replacement of provisional matrix with functional tissue. Macrophages, again, are central. They can eat dead neutrophils (the first wave of inflammatory cells), a process that helps shift signaling toward resolution. They can secrete growth factors that encourage blood vessels to mature and tissues to revascularize. They can influence how fibroblasts behave: whether they keep depositing collagen or whether they retreat and allow remodeling to proceed.

If the immune system fails to do this, a scar becomes not just a patch but a policy. It declares the region to be permanently risky, permanently in need of reinforcement. In skin that may be tolerable; in organs it can be devastating. Lung fibrosis stiffens the delicate sacs needed for gas exchange. Liver fibrosis disrupts the organ's architecture and blood flow. Kidney fibrosis compromises filtration. In each case, the immune system is involved, sometimes as an initiator, sometimes as an accomplice, sometimes as a prisoner of feedback loops that keep it returning to the scene.

The double-agent metaphor also helps explain why regeneration research can feel like negotiating with a bureaucracy. The same signals that stimulate growth and cell movement can resemble signals used by tumors. The immune system has evolved to interpret unusual proliferation with suspicion. It is not difficult to imagine why: an organism that freely allows cells to revert to a more plastic, embryonic-like behavior after injury may also be more vulnerable to cancer. The immune system's readiness to enforce order, even at the cost of scarring, may be one of the prices mammals pay for tight control of cellular behavior.

The most striking aspect of all this is that the immune system is not merely reacting to a wound; it is choosing, in a distributed and probabilistic way, the kind of tissue that will exist afterward. It decides whether the wound is treated as a temporary construction project or as a permanent fortification. It decides whether fibroblasts are asked to be scaffolders or masons. It decides whether the extracellular matrix stays soft long enough for pattern to re-emerge, or whether it hardens into a barrier that turns the wound into a historical event that the tissue can never quite forget.

That decision is written in chemistry, but it is legible in everyday experience. A cut that becomes red, hot, swollen, and increasingly painful is not merely "healing." It is a sign that demolition has intensified, perhaps because bacteria have gained a foothold. The body is choosing survival first. A wound that closes quietly, with minimal swelling, often leaves less conspicuous evidence. When clinicians talk about "excess inflammation," they are often pointing to a wound that has failed to change its mind.

Regeneration, when it occurs, requires that change of mind. Not a miraculous leap into a salamander's biology, but a disciplined transition: from danger signals to resolution signals; from tissue breakdown to tissue rebuilding; from stiffening the environment to allowing it to be remodeled. The immune system can be both guardian and saboteur because it sits at the threshold where life-saving urgency can harden into long-term limitation. The trick is not to disarm the guardian, but to teach it when the emergency is truly over.

5.3 The Epigenetic Straitjacket

A salamander can treat injury as an invitation to improvise. An adult mammal treats it as a problem to be contained, then repaired under strict rules. Part of that difference is obvious in the wound itself: the dense collagen patch, the stiffened scaffold, the immune system's lingering suspicion. The less visible part is inside the cells that remain. Even if the scar wall could be softened and the inflammatory fog cleared, most adult mammalian cells are not eager to become something else. They are specialists, and specialists keep their tools locked away.

That locking is epigenetic. The word is often used loosely, sometimes as a synonym for "gene regulation," and sometimes as a kind of biological mysticism. Here it means something concrete: chemical marks and structural arrangements that determine which stretches of DNA are accessible and which are kept shut. A cell can carry the same genetic sequence as every other cell in the body, yet behave completely differently because it reads only certain chapters of the genome. The unread chapters are not merely ignored; they are physically less reachable, packaged into tighter forms of chromatin, the DNA-protein complex that makes up chromosomes.

Epigenetics is the difference between owning a library and running a library. Every cell owns the same shelves. Only some books are on the reading tables.

During embryonic development, the library is noisy and permissive. Cells are plastic: a word that means they can still change identity and respond to patterning signals. Plasticity is not chaos; it is controlled openness. As tissues form, cells gradually commit to roles. A muscle cell becomes a muscle cell not only because it turns on muscle genes, but because it closes doors to alternative careers. A skin fibroblast does not simply fail to express neuron genes; it locks them behind layers of packaging.

Locking is useful. A body with trillions of cells needs reliability. If mature cells could easily drift into other identities, tissues would lose function and coherence. Worse, uncontrolled plasticity is one of the hallmarks of cancer. One can interpret a great deal of adult biology as a bargain: stability and tumor resistance, purchased at the price of reduced regenerative freedom.

The straitjacket metaphor has two straps. One is mechanical: the stiffness of the extracellular matrix, the physical environment that surrounds cells. The other is molecular: the tightness of chromatin and the persistence of gene regulatory "memories" that keep a cell in its lane. They reinforce each other.

Consider stiffness first. Cells are not floating in broth. They are embedded in a mesh of collagen, elastin, laminin, and other molecules that form the extracellular matrix (ECM). The ECM is not just scaffolding; it is information. Cells attach to it through receptors, pull on it, and feel how much it yields. That sensation affects how they spread, how they migrate, and which genes they express.

A classic demonstration comes from work in the 2000s by Dennis Discher's group and others, showing that stem cells can be nudged toward different fates depending on the stiffness of the surface they are grown on. Softer environments resembling brain tissue bias cells one way; stiffer environments resembling bone bias them another. The point is not that mechanics replace biochemistry, but that they join it. Physical constraints become instructions.

In adult mammalian wounds, the ECM rapidly stiffens as collagen accumulates and becomes crosslinked. Crosslinking is a chemical tightening of the collagen network, akin to adding more rivets and braces to a structure. A stiff matrix changes how cells behave. Fibroblasts on stiff substrates tend to become more activated, more contractile, more inclined to produce yet more matrix. This can lock in fibrosis. In a sense, the scar teaches the cells around it to think like scar-formers.

This mechanical teaching connects directly to epigenetics. When cells spread on a stiff surface, they generate tension in their internal skeleton (the cytoskeleton). That tension is transmitted to the nucleus through physical linkages. The nucleus is not a free-floating command center; it is a structure that can be stretched and compressed. Nuclear shape influences chromatin organization. Genes can become more or less accessible depending on how the chromatin is arranged within that physical space.

One of the reasons adult mammals struggle to regenerate is that their tissues are mechanically and epigenetically "adult." The ECM is sturdier and more crosslinked than in embryos. Many cells are already embedded in a mature architectural context that biases them toward maintaining structure rather than rewriting it. Injury adds a stiff scar patch that exaggerates these adult features. A regenerating program would need cells to loosen their commitments, move, proliferate, and respond to positional cues. A stiffened environment encourages the opposite: contract, stabilize, and stop taking risks.

The molecular strap of the straitjacket is chromatin itself. DNA is long enough that, if stretched end to end, it would measure roughly two meters in each human cell. It must fit into a nucleus only a few micrometers across. It does this by winding around proteins called histones, forming nucleosomes, which further fold into higher-order structures. Some chromatin is relatively open (euchromatin), allowing transcription machinery to access genes. Some is densely packed (heterochromatin), making genes physically harder to read.

Cells use chemical modifications to manage this packing. Histones can be tagged with methyl or acetyl groups, among other marks, which influence whether chromatin is open or closed. DNA itself can be methylated, often associated with gene silencing. These marks do not change the DNA sequence, but they can be stable over many cell divisions. Stability is the whole point: a skin cell dividing to replace itself should produce more skin cells, not decide to become cartilage on a whim.

Regeneration asks for a temporary relaxation of that stability. It asks cells to behave, briefly, less like fixed specialists and more like adaptable builders. In salamanders, many lines of evidence suggest that cells at an injury site can adopt a more plastic state, contributing to a blastema that rebuilds missing parts. In adult mammals, a comparable broad plasticity is rare and tightly constrained. When mammals do show dramatic cellular identity changes, it is often in pathological contexts: cancer, metaplasia, or chronic injury.

A historical anecdote captures how hard it was to accept that cell identity could be rewritten at all. In the 1960s, John Gurdon showed that nuclei from differentiated frog intestinal cells could, when transferred into enucleated eggs, support development of tadpoles. The differentiated nucleus, long assumed to be irrevocably committed, still contained the full genetic program for an organism. Decades later, Shinya Yamanaka demonstrated that a small set of transcription factors could reprogram adult mammalian cells into induced pluripotent stem cells, resetting them to a state resembling embryonic stem cells. These discoveries, recognized with Nobel Prizes, are often celebrated as triumphs of possibility. They are also warnings about effort. Reprogramming is not easy. It requires forcing open chromatin, rearranging epigenetic marks, and pushing cells through unstable intermediate states that can be error-prone.

In a wound, the body is not trying to win a Nobel Prize. It is trying to prevent infection and maintain integrity. So it relies on programs that are robust and safe. Robustness tends to mean "hard to reprogram."

Yet wounds do involve epigenetic change. A repair response is not simply a chemical reflex; it requires turning on genes that are not normally active in intact tissue. The epidermis at the wound edge, for example, must switch from a barrier-maintenance mode to a migration-and-closure mode. Research in the 2010s showed that murine wound epidermis undergoes regulated changes in Polycomb-mediated repression. Polycomb group proteins help keep developmental genes silenced by placing repressive marks such as H3K27me3 on histones. After injury, parts of this repression can be eased, allowing a gene program appropriate for repair to activate.

That observation carries two implications. First, epigenetic remodeling is not exotic; it is part of ordinary healing. Second, the remodeling is selective. The genome is not thrown open wholesale. The cell unlocks only the compartments it believes it needs for the emergency at hand. In adult mammals, the emergency is closure and stabilization, not full pattern restoration. The epigenetic changes that occur tend to support rapid migration, inflammation response, and barrier repair. They do not necessarily restore the broad developmental plasticity required for complex regeneration.

The epigenetic straitjacket is also tightened by memory. Cells can remember prior inflammatory or fibrotic experiences in ways that persist long after the original trigger is gone. In chronic fibrotic diseases, fibroblasts can become locked into self-sustaining loops where they keep producing extracellular matrix and pro-fibrotic signals even when the initiating inflammation has subsided. This has been described as an enhancer-level memory: enhancer regions of DNA (regulatory switches that control gene expression) become persistently activated, making the fibrotic program easier to reinitiate and harder to shut down.

A concrete example comes from systemic sclerosis, a disease marked by widespread fibrosis. Studies have implicated persistent activation of regulatory elements associated with TGFB2, part of the TGF-β (transforming growth factor beta) signaling family that is central to fibrotic activation. The important point is not the gene name but the logic: fibroblasts can become the kind of cells that *expect* to make scar. They do not need constant external shouting from immune cells; their own chromatin has been tuned to keep the volume up.

Another mechanistic window into this tuning involves BRD4, a protein that binds acetylated histones and helps drive transcription at key regulatory sites, including so-called super-enhancers. In cardiac and other fibroblast systems, BRD4 has been shown to retarget during activation, helping to establish or maintain pro-fibrotic gene expression programs. Stress signaling pathways such as p38 MAPK can feed into these chromatin-level changes, linking external injury cues to durable changes in transcriptional control. This is epigenetics as bureaucracy: once the new forms are filed and stamped, reversing the decision requires more than telling the office to calm down.

These molecular details matter for regeneration because regeneration is not simply a matter of making cells divide. It is a matter of making cells divide *and* differentiate in the right pattern, while rebuilding an environment that supports that pattern. A wound that has moved into a fibrotic, stiffened, epigenetically reinforced state is like a city that responded to an earthquake by pouring concrete over the ruins and declaring the neighborhood permanently industrial. One can still build there, but it will not become what it was.

The Hippo-YAP pathway provides a particularly intuitive bridge between mechanics and gene regulation. YAP (Yes-associated protein) is a transcriptional co-activator that responds to mechanical cues; when it enters the nucleus, it can influence gene expression related to growth, survival, and cell fate. Work comparing spiny mice and ordinary mice has implicated differences in myofibroblast behavior and YAP-related programs in scar-free versus scar-forming outcomes. Again, the names are less important than the theme: the wound environment can push a cell toward a durable program, and that program can be stabilized through transcriptional and epigenetic mechanisms.

At the bedside, the epigenetic straitjacket shows up as a familiar frustration. Surgeons can cut and suture with exquisite precision, but they cannot simply request that adult skin rebuild with hair follicles aligned and sweat glands restored. Cardiologists can reopen blocked arteries, but they cannot easily persuade the adult heart to replace a dead patch of muscle with new muscle rather than a collagen scar. Even when stem cells are introduced experimentally, they land in a landscape that tells them, mechanically and epigenetically, what kind of tissue is permitted.

There is also a more intimate reason to care. Scars are not merely marks; they can restrict movement, cause chronic pain, and reshape lives. Severe burns can lead to contractures where scar tissue tightens over joints. In organs, fibrosis can be quietly progressive until it crosses a threshold into failure. The biology that prevents infection and bleeding in the short term can become a slow theft of function in the long term.

The phrase "turning back the clock" is often used to describe regenerative hopes: make adult cells youthful again, make tissues behave more like embryos. The straitjacket metaphor clarifies why that is difficult. Youthful is not just a matter of having more growth factors. It is a matter of having chromatin that is more permissive, an extracellular matrix that is less stiff and less crosslinked, and a regulatory system that tolerates temporary ambiguity in cell identity without tipping into cancer.

Nature's solutions to this are revealing. The fetus can heal with less scarring in part because the inflammatory environment is different, but also because tissues are mechanically softer and cells are still in the process of committing to identities. The salamander can regenerate in part because its wound environment stays permissive longer, and because its cells and immune system cooperate to prevent early consolidation of a fibrotic wall. Adult mammals, in contrast, are optimized for stability. Their cells are built to remember what they are.

The most honest way to describe the obstacle is not that adult mammals lack regenerative genes. The genes are there, in the same genome that once built an embryo. The obstacle is access and permission. The genome is a locked workshop; the ECM is a rigid scaffolding; the immune system is a vigilant foreman wary of uncontrolled growth. Injury can temporarily loosen a few locks, just enough to close the breach. Full regeneration would require a broader amnesty: a controlled interval in which cells are allowed to become plastic, move through a softer matrix, and rebuild architecture without being forced back into fibrotic compliance.

When people imagine regeneration, they often picture a dramatic moment: a limb bud forming, a heart muscle regrowing, skin becoming seamless. In mammals, the more decisive moments are quieter. They happen in the first days after injury, when the matrix stiffens and immune signals peak, and when the nucleus of a fibroblast decides which doors in its library will remain closed. A scar is not only a patch of collagen on the outside; it is a set of locked chromatin states on the inside, maintained by a tissue that has learned, in the most literal molecular sense, to prefer certainty over possibility.

Chapter 6

The Faustian Bargain

We often view our inability to regrow limbs as a biological defect, yet evolution rarely retains traits without purpose. This chapter argues that our frailty is actually a shield-a necessary sacrifice made to protect long-lived, warm-blooded bodies from the ravages of uncontrolled cell growth and infection. We explore the hypothesis that humanity traded the miracle of regeneration for the stability required to survive a complex, high-energy existence.

6.1 The Cancer Calculus

A salamander that loses a limb does not merely heal the wound. It does something closer to time travel. Skin cells, connective tissue cells, even muscle-associated cells near the cut edge loosen their identities, step back from their settled adult jobs, and re-enter a state of productive uncertainty. They proliferate, migrate, and listen to positional instructions until a structure reappears with the right proportions: wrist where a wrist should be, digits where digits belong. The remarkable part is not only that this happens, but that it happens without the animal reliably dying of cancer.

To a mammal, that same recipe reads like a crime report.

Cancer begins, in its simplest description, when a cell acquires permission to ignore the social contract of multicellular life. It divides when it should not, refuses to die when it should, squats in tissues where it does not belong, and recruits support. The early steps of that process are not exotic. They look uncomfortably like what any organism must do to build new tissue: loosen constraints, re-enter the cell cycle, rewire gene expression, and sometimes adopt a more embryonic program. *Regeneration* and *tumorigenesis*—the birth of a tumor—share verbs even if their sentences end differently.

This is the calculus: the more willing a body is to let its cells become plastic—able to change identity, to proliferate quickly, to move and remodel—the more it must trust that those same cells will stop on command. If the stopping mechanism is imperfect, plasticity becomes peril. If the stopping mechanism is strict, plasticity becomes scarce.

The mammalian body, as a rule, chooses strict.

A central figure in that choice is a protein called p53, often nicknamed the "guardian of the genome." The nickname sounds like marketing, but it captures the job. When DNA is damaged, when chromosomes misbehave during division, when growth signals surge in ways that smell wrong, p53 can force a cell to pause, attempt repair, or self-destruct through apoptosis (a controlled cellular suicide). p53 is not the only checkpoint, but it is one of the most consequential, and it sits at an intersection between two needs that rarely agree: make more cells, and keep those cells safe.

It helps to imagine a tissue as a city. Cell division is construction. DNA replication is the copying of blueprints. Every act of copying introduces the possibility of error; every day of construction raises the chance that someone cuts corners. If your city expects to rebuild whole neighborhoods after a storm, it must tolerate a lot of construction. If your city is terrified of structural collapse, it will demand inspections at every stage. Those inspections save lives, but they also slow rebuilding. Mammals are the city that grew up with strict inspectors.

One reason is simply arithmetic. Cancer risk is shaped by the number of cells in a body and the number of times those cells divide over a lifetime. A mouse is small and usually short-lived; a whale is vast and can be long-lived. If each cell division carries some chance of acquiring a cancer-promoting mutation, then more cells and more divisions ought to mean more cancer. This expectation, long discussed under the banner of "Peto's paradox," collides with the observation that large animals are not, in any naive sense, cancer factories. Elephants do not commonly die of malignancies at the rate one might predict from their size alone. This does not mean big animals are immune; it means evolution has had to invest in countermeasures.

Those countermeasures are revealing, because they show what it costs to keep a large, long-lived body stable. In 2015, Lisa Abegglen, Aleah Caulin, and colleagues drew attention to one striking feature of elephant genomes: multiple copies of *TP53*, the gene that encodes p53. Humans have one canonical copy; elephants have many more. More copies can mean more p53 activity, more readiness to trigger apoptosis, more hair-trigger caution. Later work identified an additional twist: a "zombie" gene, *LIF6*, a reawakened pseudogene that can be activated by p53 and help push damaged cells toward death. The evolutionary logic is blunt. When you have a lot of cells and a lot of years, you cannot afford to be forgiving.

But forgiveness is exactly what regeneration asks for—at least temporarily. When an adult tissue is wounded, cells around the wound must interpret damage signals, change their behavior, and sometimes adopt identities that are not their final form. In salamander limb regeneration, a structure called the blastema forms: a mound of proliferating, progenitor-like cells that will rebuild the missing parts. A blastema is not a tumor. It is organized, patterned, and ultimately self-limiting. It emerges in a context rich in cues that say, in effect, "build a forearm, stop at the wrist, add four fingers, and quit." Yet to get there, cells must do things mammalian cells are trained to treat as red flags: dedifferentiate (move away from a fixed identity), proliferate rapidly, and remodel the extracellular matrix (the meshwork of proteins and sugars that gives tissue its structure).

In mammals, the moment an adult cell behaves as though it is slipping its identity and re-entering a more embryonic mode, p53 and its allies are more likely to intervene. That intervention can be life-saving in one context and thwarting in another. The same brake that prevents a precancerous cell from becoming a malignancy can also prevent a wounded tissue from building a blastema. The calculus does not say regeneration *is* cancer. It says regeneration *resembles* the early steps of cancer closely enough that the anti-cancer system has a hard time telling friend from foe.

A historical anecdote makes the tension vivid. In the eighteenth century, the naturalist Lazzaro Spallanzani became famous for careful experiments on animals that regrow parts. He described salamanders regenerating limbs and tails with an almost unsettling calmness, as if the body contained a hidden workshop that could be reopened. Readers in Europe were captivated, but also suspicious: if an animal can make a limb anew, why can it not grow monstrous extra parts endlessly? The implicit fear was that regeneration might be uncontrolled growth by another name. Spallanzani did not have molecular tools, but his question was the right one. Regeneration requires growth; life depends on controlling growth.

Modern biology has turned that philosophical worry into mechanistic detail. Cells have layers of safeguards against abnormal proliferation. Some are like speed limits (signals that slow division). Some are like seatbelts (repair systems that fix DNA damage). Some are like airbags (apoptosis, which destroys a cell that has become too risky). p53 is a coordinator: a dispatcher deciding whether the cell should brake, repair, or exit the road entirely.

Now imagine a cut limb stump trying to rebuild. The local environment becomes a storm of signals: reactive oxygen species from injury, inflammatory molecules, disrupted blood flow, and mechanical changes in the tissue scaffold. DNA damage is more likely under stress. Yet regeneration needs cells to divide anyway. That is a paradox in miniature: to regenerate, cells must accept a period in which mistakes are more likely, while also ensuring those mistakes do not propagate into cancer.

Some species appear to solve this by timing. The regenerative program can include a carefully tuned relaxation of certain checkpoints during the early proliferative phase, followed by a reassertion as patterning and differentiation proceed. The body permits a controlled flirtation with danger, then insists on commitment. Mammalian wound healing, in contrast, behaves like a cautious system designed to avoid flirting at all. It pushes quickly toward closure and stability, even if that stability is imperfect.

This is why the word *plasticity* matters. Plasticity is the capacity of a cell to change what it is. An adult skin cell is usually not supposed to become a cartilage cell; a muscle-associated cell is usually not supposed to behave like a generic progenitor. Salamanders can coax adult cells into something closer to a developmental state. Mammals can do this in narrow contexts—think of the liver's partial regenerative capacity, or stem-cell-driven renewal in blood and intestine—but broad, limb-style plasticity is unusual. Not because the genes for building limbs are missing. Mammals build limbs perfectly well in the embryo. The issue is access: whether adult cells are allowed to reopen those programs without being stopped as potential threats.

The immune system adds another layer to the calculus. A vigilant immune system is good at detecting the abnormal—cells displaying strange proteins, stressed cells signaling distress, cells that do not fit the tissue's norms. That vigilance is valuable against viruses and cancers. It can also interpret regenerating tissue as suspicious, especially if cells are changing identity and proliferating in a chaotic chemical landscape. Inflammation is not simply collateral damage; it is a policing strategy. But policing can be heavy-handed.

The mammalian strategy can be summarized as: *do not let adult cells become too adventurous.* When they must become adventurous, force them to prove they are safe at every step. The price is paid at the wound site.

A useful way to see the similarity between regeneration and cancer is to lay their early steps side by side. Both begin with injury or insult (a cut, a toxin, a DNA-damaging event). Both involve stress signals and changes to the extracellular matrix. Both can involve cells re-entering the cell cycle. Both can involve a partial loss of differentiated identity. The crucial difference is governance. Regeneration is a construction project with an architect, a schedule, and an end point. Cancer is construction without permits.

The tragedy for mammals is that their permit office is so strict that it sometimes refuses legitimate rebuilds.

Parallel early steps: blastema formation vs. tumor initiation

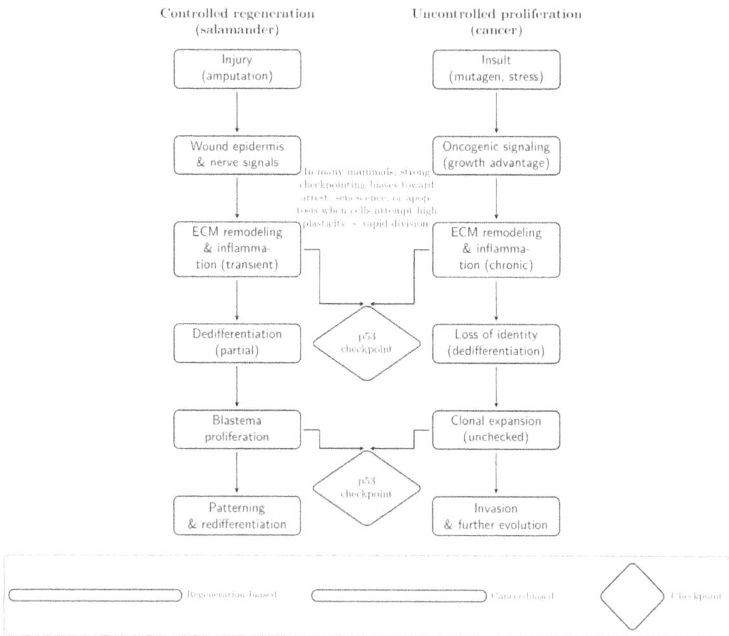

The diagram is not meant to claim that every regenerating cell is one step from malignancy, or that every cancer imitates a blastema. It is meant to show why the distinction is hard to enforce using simple rules. If your rule is "cells should not dedifferentiate" or "cells should not proliferate rapidly under stress," then you will prevent many cancers—but you will also prevent the most dramatic forms of regeneration.

That helps explain why some of the most regeneration-competent animals seem to have evolved solutions that are not merely permissive, but *clever*. They do not throw away tumor suppression. They choreograph it. If p53 is too lax, tumors would bloom. If it is too strict, the blastema never forms. Evolution, in those lineages, has found a dynamic compromise.

Mammals, meanwhile, have built a different compromise suited to a different life. Warm-bloodedness brings high metabolic rates, constant oxygen flux, and with it a steady background of molecular wear. A mammal's tissues are also often designed for performance: rapid muscle response, high-energy brains, efficient immune defense. In such a body, the cost of a single renegade clone can be catastrophic. Natural selection tends to prize a system that prevents early death from cancer over one that permits occasional perfect reconstruction of a lost limb. The arithmetic of reproductive success is unforgiving: an animal that survives to reproduce passes on its genes, even if it carries an invisible debt of scarring and diminished plasticity.

The calculus becomes even more intimate when one considers that tumor suppression is not merely a barrier erected *against* abnormal growth. It is woven into normal development and maintenance. Checkpoints regulate stem cells, constrain tissue architecture, and help maintain differentiation—keeping cells what they are supposed to be. Those constraints make adult tissues reliable. Reliability is not glamorous, but it is the difference between an organ that functions for decades and one that drifts into chaos.

There is, however, a haunting flip side. If mammalian cells are trained to interpret plasticity as danger, then regenerative medicine is asking them to violate their own safety culture. Every approach that aims to regrow tissue—coaxing cells to dedifferentiate, transplanting stem cells, reprogramming adult cells—must contend with the surveillance apparatus. Not as an enemy, but as an anxious guardian that has been right, statistically, most of the time. The challenge is not merely to make cells grow. It is to make them grow *under governance*.

That governance may be biochemical (timed checkpoints), ecological (an immune environment that supports rebuilding rather than panic), or architectural (extracellular matrix cues that channel proliferation into form). In animals that regenerate well, those elements align. In mammals, the alignment tends to favor containment.

If this feels like a pessimistic story, it is also a story of engineering excellence. The mammalian body is astonishing at preventing catastrophe given the sheer number of risky events it faces. Every day, cells divide, DNA is copied, and errors occur. Most are corrected. Some are not. Yet most humans do not develop cancer in childhood, when cell division is rapid; most do not develop it in early adulthood, when life is often at its most metabolically intense. The safety systems work, and they work by limiting the very behaviors that regeneration would exploit.

The bargain is written into every scar and every tumor that never appears: a body that prefers to patch and stabilize, to preserve order even at the cost of lost possibilities.

6.2 The Erosion of Time

A child's scraped knee can look almost miraculous. The skin knits with a speed that feels automatic, the redness fades, and within days the injury becomes a faint memory. The same scrape on an older adult can linger: slower closure, more persistent inflammation, a darker mark that stays for months. The difference is not only a matter of circulation or "slower metabolism." It reflects a deep change in how cells interpret instructions. Youth is not just time with fewer mistakes; it is time when the body is more willing to rewrite itself.

The word that captures that willingness is *plasticity*. In a young organism, many cells remain capable of changing their behavior dramatically in response to signals. With age, that flexibility narrows. Some of that narrowing is protective. A body that has lived longer has had more opportunities to accumulate mutations, more exposures to viruses and toxins, more rounds of cell division. The anti-cancer brakes described earlier do not retire; if anything, they become more suspicious. Yet suspicion has a cost. A tissue that treats every call to proliferate as a potential coup will, sooner or later, begin to refuse legitimate repair work.

Two forces drive this refusal. One is the gradual tightening of the genome's "operating system"—the epigenetic state that determines which genes are easy to turn on and which are hard to access. The other is the slow change in tissue environment—the inflammatory background noise, the accumulation of senescent cells, the stiffening of scaffolds—that makes delicate regenerative conversations harder to hear.

Epigenetics is often described with metaphors that risk sounding mystical, but the underlying idea is concrete. Every cell in your body carries essentially the same DNA sequence, yet a neuron behaves differently from a liver cell because different genes are active. Those activity patterns are regulated by chemical marks on DNA and histone proteins, and by the packing of DNA into chromatin. Tightly packed chromatin is less accessible: genes there are harder to activate. Loosely packed chromatin is more accessible: genes can be read and used. Epigenetic marks are not the genetic text; they are the punctuation and page layout that decide what can be read at a given moment.

Early in development, the genome is comparatively open and negotiable. As tissues mature, chromatin landscapes become more specialized, more locked-in. That is part of what makes adult organs stable. A heart cell that casually decides to behave like a dividing embryonic cell is not charming; it is lethal. The same principle that keeps identity stable also makes identity harder to change when change would be useful—after injury, during repair, in the attempt to rebuild complex structures.

One of the most influential ways to picture this is through the developmental biologist Conrad Waddington's famous image from the mid-twentieth century: an "epigenetic landscape" in which a marble rolls down a hill into valleys. Each valley is a cell fate. In early life the marble can still be nudged into different valleys; later it sits deep in one groove. Waddington's drawing was a metaphor before epigenetic mechanisms were well understood, but it captured something that modern chromatin biology has made physical. With age, many cell identities become deeper grooves.

If this were only about a tidy hardening of identity, it might not sound so harmful. The real problem is that injury repair often requires controlled exceptions. A wounded tissue needs some cells to proliferate, some to migrate, some to adopt temporary roles. It needs the genome to reopen certain developmental programs briefly, then close them again. Aging makes both the reopening and the reclosing less precise.

A vivid demonstration of reversibility—of the idea that cell identity can be rewritten—arrived in 2006, when Shinya Yamanaka showed that introducing a small set of genes could reprogram adult mouse cells back into an embryonic-like state. The induced pluripotent stem cell, or iPSC, was a conceptual earthquake: it proved that adult identity is maintained actively, not simply cemented forever. Yet the experiment also illustrated the danger. The very factors that restore youthful plasticity can promote tumor formation if mismanaged. Reprogramming and cancer are uncomfortable neighbors. In the clinic, that neighborhood has forced a cautious pace: the promise of rebuilding tissue is always shadowed by the risk of growing the wrong thing.

Reprogramming also exposed a quieter truth: older cells resist. They accumulate epigenetic "memory" that makes them harder to reset cleanly. Some of that resistance arises from DNA damage checkpoints. Some comes from chromatin itself. In practice, the older the donor, the more stubborn the cells can be, and the more likely they are to acquire abnormalities during extensive manipulation. The lesson for regeneration is sobering. If even a laboratory can struggle to persuade aged cells to become flexible under tightly controlled conditions, a living tissue—messy, inflamed, and constrained by survival priorities—has an even harder task.

Aging is not a single process but a pile-up of processes, and one of the most consequential is *cellular senescence*. A senescent cell is alive but permanently withdrawn from the cell cycle. It no longer divides. Senescence can be triggered by telomere shortening (the gradual loss of protective chromosome ends with repeated division), by DNA damage, by oncogenic stress, and by other disruptions. From the perspective of cancer prevention, senescence is a triumph: a potentially dangerous cell is forced into retirement.

From the perspective of regeneration, senescence is ambiguous, and that ambiguity is part of why evolution has tolerated it.

For years, senescent cells were framed mainly as villains of aging. They accumulate in tissues with time and secrete a cocktail of inflammatory and tissue-remodeling molecules known as the senescence-associated secretory phenotype (SASP). This secretion can disrupt neighboring cells, promote chronic inflammation, and alter tissue structure. In aged tissues, clusters of senescent cells can create a microenvironment that is both noisy and corrosive—bad for stem cells, bad for repair.

Then an unexpected twist emerged. Judith Campisi and colleagues had long emphasized senescence as a tumor-suppressive program; later work, including striking studies in the 2010s, showed that senescent cells can play a helpful, time-limited role in wound healing. After injury, certain fibroblasts and endothelial cells can enter senescence transiently and secrete factors that promote closure, including signals that encourage myofibroblast differentiation, the contractile behavior that pulls wounds together. In other words, senescence is not merely an end-stage pathology. It can be deployed as a tool: halt risky proliferation, but still contribute to repair through signaling.

The problem is that tools become trash when they are not cleared away. In younger tissues, transient senescent cells are more effectively removed, often by immune surveillance. With age, clearance becomes less efficient, and senescent cells persist. Their SASP, once a short-lived flare, becomes a chronic smolder. The tissue environment begins to resemble an injury that never quite resolves.

That chronicity has a name: *inflammaging*, a low-grade, long-term inflammatory state that tends to rise with age. It is not the same as an acute infection—fever, swelling, and dramatic immune mobilization. It is subtler: slightly elevated inflammatory cytokines, altered immune cell composition, and a steady background of signals that say "something is wrong." That background has consequences for regeneration. Many regenerative processes depend on precisely timed immune activity: an early cleanup phase, a transition to rebuilding signals, then a return to calm. When the immune system is already humming with generalized irritation, the transition becomes difficult. Repair signals are drowned in static.

Aging also reshapes the extracellular matrix (ECM), the physical and biochemical scaffold that surrounds cells. Collagen cross-linking increases, tissues stiffen, and the ECM becomes less like a flexible workspace and more like old drywall. Cells sense stiffness. They respond to it by changing gene expression, sometimes by becoming more fibrotic—producing more scar-like matrix—rather than more regenerative. A stiff environment can push stem cells toward exhaustion or inappropriate differentiation. It can also amplify inflammatory signaling, creating a feedback loop: inflammation changes matrix, matrix reinforces inflammation.

The consequences show up in places where people feel them most. Muscle repair after injury becomes less complete. Tendons heal slowly and often with reduced elasticity. Skin becomes thinner and more fragile. Even the brain's limited capacity for certain forms of plasticity declines, in part because supportive environments shift. Aging is not only internal to the cell; it is in the neighborhood.

At the molecular level, one can think of aging as a rising cost of cell division. Every division risks copying errors. Every division shortens telomeres in many cell types. Every division tests repair systems. If an organism wants to live a long time, it must either accept rising cancer risk or invest in systems that reduce that risk. Some long-lived species appear to invest heavily in genome maintenance. Work on bowhead whale cells, for example, has suggested unusually robust DNA double-strand break repair and lower mutation accumulation. That kind of investment is appealing: fewer mutations means fewer triggers for p53 alarms, fewer cells pushed into senescence, fewer reasons to clamp down on proliferation.

Yet high-fidelity repair and strict checkpoints can also reinforce epigenetic rigidity. A cell that is highly committed to maintaining its state will treat dramatic transcriptional rewiring—turning one gene program off and another on—as suspicious. In humans, even modest attempts to push adult cells into unfamiliar behaviors can activate stress responses. The body is not merely protecting against tumors; it is protecting against identity drift, mispatterning, and organ failure. Age increases the stakes because the buffer capacity is lower: fewer reserve stem cells, less redundancy, more accumulated wear.

This helps explain a poignant pattern seen across many tissues: stem cells remain, but their output changes. In the blood system, aged hematopoietic stem cells often skew toward producing myeloid cells over lymphoid cells, reshaping immune function. In muscle, satellite cells can become less responsive and more prone to adopting fibrotic fates under inflammatory cues. In skin, stem cell pools and their niches change. These are not simply stem cells "running out." They are stem cells living under a different governance regime, in a different epigenetic and inflammatory climate.

An anecdote from clinical history captures the human stakes without requiring any molecular details. Before antibiotics were widespread, a deep wound was a gamble. Quick closure could mean survival; a slow, open wound could mean infection and death. Even today, chronic wounds in older adults—diabetic ulcers, pressure sores—can become life-altering, leading to pain, repeated hospital visits, and amputations. The body's bias toward rapid sealing and strict control is not an aesthetic preference; it is a survival strategy that has been selected under the constant threat of infection. Aging erodes the ability to execute that strategy cleanly. The same immune responses and fibrotic tendencies that once saved lives can, in an older body, overshoot into chronic inflammation and scar-heavy repair.

If aging were merely a matter of time, one might expect all tissues to degrade uniformly. They do not. Some organs maintain function surprisingly well. The liver often retains an impressive capacity to restore mass after injury. Blood replenishment continues. Intestinal lining renews rapidly. These tissues share a feature: they already live with high turnover and robust stem cell activity. They are designed to proliferate safely, with built-in architecture that channels renewal. Even there, however, age increases cancer incidence and decreases regenerative precision. The liver can regrow mass but not always perfect structure; fibrosis can develop with chronic injury; cancer risk rises with decades of replication.

The tissues that fail most dramatically are those that require not just replacement of cells, but the reconstruction of complex, low-turnover architecture—precisely the sort of challenge posed by limbs, heart muscle, and certain nervous system structures. Those tissues rely heavily on stability. With age, stability becomes a fortress.

One can see the fortress in chromatin itself. As cells age, patterns of DNA methylation shift in ways that are so consistent that they can be used to estimate biological age, the basis of "epigenetic clocks." These clocks are not merely curiosities. They suggest that aging includes an organized drift of regulatory state. Whether that drift is a cause of aging or a record of it is debated, but either way it marks the genome as less like an open library and more like an archive with increasing restrictions.

The metaphor of an archive is useful because it captures both the promise and the frustration. The programs for regeneration are not entirely absent in mammals; embryos run them with confidence. The adult genome retains the instructions, but access is controlled. Aging adds locks, and the keys become harder to find. Even when a signal manages to open a cabinet, the surrounding room is louder, more crowded, more brittle. Repair becomes possible in principle but difficult in practice.

None of this means that youth is a free-for-all. Young tissues also use checkpoints; they also deploy senescence; they also balance proliferation against safety. The difference is that young tissues tend to resolve crises. Inflammation rises and falls. Senescent cells appear and are cleared. Chromatin states remain firm enough to preserve identity but flexible enough to allow controlled remodeling. Aging is what happens when resolution becomes rare: when the end of the story is delayed, and the body keeps living in the middle of an alarm.

Regeneration, at its most ambitious, requires quiet moments of coordinated risk: cells must proliferate, change identity, and move, all while being supervised rather than strangled. Aging replaces supervision with suspicion, and suspicion with exhaustion. A wound, even a small one, becomes a negotiation with a bureaucracy that has learned too many hard lessons and has too few clerks left to process exceptions.

6.3 Evolution's Ledger

A perfect limb is a triumph, but in the wild perfection can be a luxury good. Evolution does not reward beauty, symmetry, or completeness unless those traits pay rent in survival and reproduction. A wound is not a philosophical problem; it is blood loss, exposure, pain, and vulnerability. The body's first question is rarely "Can I rebuild this?" It is "Can I keep the inside in, and the outside out, long enough to see tomorrow?"

That question has a peculiar cruelty. If you imagine a salamander calmly assembling a new forelimb, it is easy to read mammalian scarring as a failure, a regrettable compromise, an ancient defect. A better way to see it is as a strategy that won, again and again, in the arithmetic of natural selection. Scars are not merely evidence that regeneration did not happen. They are evidence that a wound was treated as an emergency breach.

A breach has enemies. Predators are obvious. Infection is often deadlier.

Long before anyone knew what bacteria were, the clinical reality of wounds taught the lesson. A deep cut could fester into gangrene. Childbirth injuries could become fatal. A small puncture could produce swelling, fever, delirium, and death. The nineteenth-century surgeon Joseph Lister is remembered because he offered a way out: antiseptic surgery. In the 1860s, he began applying carbolic acid to surgical wounds and instruments, dramatically lowering infection rates. The historical point is not simply that microbes exist. It is that for most of mammalian history, an open wound was an invitation for invisible attackers. When the default world includes pathogens ready to colonize damaged tissue, speed becomes a virtue. Tight sealing and rapid closure can matter more than architectural perfection.

Regeneration, in its most impressive forms, is slow. It asks for time: days to form a blastema, then weeks to pattern and rebuild. That is a long interval to remain compromised. The interval is not merely waiting; it is an active period in which tissues are remodeling, cells are proliferating, nerves are regrowing, blood vessels are reorganizing. During that time, the injured animal must still eat, evade predators, compete for mates, maintain body temperature, and avoid sepsis.

The evolutionary ledger begins with a simple trade: a fast, rough repair that restores the barrier quickly versus a slow, exquisite rebuild that restores the original form. Mammals generally pay for speed.

The core mechanism of that speed is *fibrosis*, the process that produces scar tissue. Fibrosis is often described as an overreaction, but in a fresh wound it is a construction choice: deposit collagen, contract the wound, stabilize the area, and create a patch that resists tearing. Scar tissue is not identical to the original tissue. It is frequently less elastic, less functional, less integrated with nerves and specialized structures. Yet it is strong. It closes holes.

The shift from regeneration to fibrosis does not happen only at the level of "genes for regrowth." It happens at decision points in wound healing, where immune signals, mechanical forces, and extracellular matrix (ECM) composition push the tissue toward rebuilding or toward sealing.

The immune system is central to this decision. In many mammals, injury triggers an inflammatory response that is rapid and intense: immune cells flood in, debris is cleared, and antimicrobial defenses are deployed. That response is lifesaving in a dirty world. It is also disruptive. Inflammation generates reactive molecules, proteases, and cytokines that can damage delicate structures and alter signaling gradients. It creates a chemical storm. Regeneration, by contrast, often requires a more carefully staged immune choreography, with early cleanup followed by signals that promote tissue patterning rather than persistent alarm.

There is a revealing developmental clue. Mammalian fetuses can heal skin wounds with minimal scarring in early gestation, a phenomenon that has fascinated surgeons and developmental biologists for decades. As development proceeds, the same organism shifts toward the adult pattern: more inflammation, more fibrosis, more scarring. The timing of the shift overlaps with maturation of immune competence. This does not prove that immunity *causes* the loss of regeneration, but it supports a plausible bargain: a stronger, faster inflammatory defense may come with reduced tolerance for the slow cellular plasticity and remodeling that regeneration demands.

The bargain becomes even clearer when one considers mechanics. Mammalian skin wounds tend to close partly by *contraction*. Specialized cells called myofibroblasts form within the wound bed, attach to the ECM, and pull the edges inward like drawstrings. Contraction reduces wound area quickly. It also distorts local geometry, changes tension patterns, and can encourage dense collagen deposition. In some experimental models, reducing contraction—by altering mechanical tension or interfering with myofibroblast activity—can shift healing toward more regenerative outcomes, with better tissue architecture and less scarring. The implication is not that mammals *cannot* regenerate, but that their default mechanical toolkit pushes them toward closing the breach first and asking aesthetic questions later.

Why would mammals, of all animals, lean so hard into speed and contraction? Metabolism offers part of the answer.

Mammals are warm-blooded. Maintaining a stable internal temperature is energetically expensive. High metabolism supports rapid movement, sustained activity, complex brains, and the ability to inhabit diverse climates. It also raises the price of downtime. A small mammal that cannot forage effectively for weeks while a limb regenerates is under immediate energetic stress. Even a larger mammal faces the problem that a prolonged injury is not merely a local defect; it can destabilize the entire energy budget. Muscle wasting, impaired hunting or grazing, and reduced ability to escape predators compound quickly.

Amphibians and many fish operate under different constraints. They are often ectothermic, relying on external heat sources. Their metabolic rates can be lower, and their lifestyles may tolerate periods of reduced activity. Many of them also live in aquatic environments where wounds are less prone to desiccation, and where the mechanical context of healing differs. Water does not eliminate infection, but it changes the physical and physiological demands of wound closure. The point is not that water is safe; it is that the environment shapes which healing strategy pays off.

Ecology sharpens this further. A bleeding wound is a signal. In terrestrial settings, blood and odor can attract predators and scavengers. A limping mammal is easier to catch. An animal that remains injured for weeks while building a perfect replacement structure may not live long enough to finish the project. A fast scar, even a disabling one, may allow the animal to survive the immediate danger.

This is where an idea from evolutionary biology, *antagonistic pleiotropy*, turns the ledger into a narrative. The phrase sounds technical, but its meaning is intuitive: a single trait can have both benefits and costs, and if the benefits appear early in life—when reproduction is likely—natural selection can favor the trait even if it causes harm later. Strong inflammatory responses, rapid fibrosis, and vigilant tumor suppression can be enormously beneficial for surviving infections and avoiding cancer during the reproductive years. The price may be paid later as chronic inflammation, fibrotic diseases, and reduced regenerative capacity. Evolution counts babies, not birthdays.

A poignant example comes from human biology itself. The fetus can heal without scars early on, and then loses that ability before birth. If scarless healing were simply "better," it is hard to explain why it would be discarded so quickly. The more plausible interpretation is that scarless healing is better *under fetal conditions*: a protected environment, low pathogen exposure, different immune dynamics, and different mechanical demands. Once the organism must defend itself in a microbe-rich world, the priorities reorder. A wound must become a barrier quickly, even if the barrier is imperfect.

Scars also support another priority that mammals tend to value: stability of tissue identity. Complex regeneration requires cells to dedifferentiate, proliferate, and then redifferentiate with exquisite positional control. That kind of control is easier in animals that can tolerate a temporary loosening of cellular identity without triggering the full machinery of cancer prevention. Mammals, especially long-lived ones, have less margin for such loosening. If the anti-cancer system described earlier is a strict inspector, fibrosis is the emergency contractor who arrives with plywood and nails, boards up the broken window, and leaves a functional but ugly patch. It is not artistry; it is triage.

There is another, less obvious reason that scarring can win: tissues can remain *functional enough* with scars. Mammalian bodies are built with redundancies and compensations. A scarred skin patch still blocks pathogens. A fibrotic liver can still perform many tasks until scarring becomes extensive. A healed tendon may be stiffer but can transmit force. Imperfection is tolerable if it restores adequate performance quickly. Natural selection often prefers "good enough, fast" over "perfect, slow."

This does not mean fibrosis is always adaptive. Chronic fibrosis can be deadly: pulmonary fibrosis stiffens lungs; cardiac fibrosis compromises heart function; liver cirrhosis destroys architecture; kidney fibrosis leads toward failure. Those pathologies are part of the later-life bill. But the ledger that matters most is written earlier, under the pressures of infection, predation, and reproduction.

The subtler question is why the mammalian healing response is so difficult to redirect once it begins. Part of the answer lies in feedback loops. Inflammation recruits fibroblasts and promotes myofibroblast differentiation; myofibroblasts contract and deposit collagen; collagen and stiffness alter cell signaling and can prolong inflammatory programs; persistent inflammation sustains fibrotic signaling. Once the wound environment crosses certain thresholds—too stiff, too inflamed, too contractile—the path back toward regeneration becomes narrow. A salamander's wound environment is often described as permissive for blastema formation, with ECM components and immune signaling that support cell migration and patterning. Mammalian wounds, by default, become a battleground fortified with collagen.

There are small windows in which mammals show hints of other options. The MRL mouse strain, for example, has been reported to close ear punches with unusually good tissue restoration compared to typical laboratory strains, reigniting interest in whether mammalian regeneration is truly absent or merely suppressed. Such cases do not overturn the general pattern. They are reminders that the machinery is not binary. Mammals are not missing the tools so much as they are using them under a different set of rules.

Those rules may be mechanical as much as molecular. The act of wound contraction is a physical decision: pull edges together now or leave them open while rebuilding proceeds. Contraction reduces the exposed surface area and the time pathogens have to invade. It also reduces the time an animal spends impaired. But contraction reorganizes tissue rapidly, sometimes at the expense of fine patterning. A body that commits early to contraction is, in effect, betting against complex regeneration.

It is tempting to frame the story as if mammals chose scarring and therefore lost regeneration. In reality, there was no committee meeting. The shift likely emerged from many small advantages layered over millions of years: slightly faster closure, slightly better infection resistance, slightly fewer lethal cancers, slightly improved survival after injury. Each advantage would be amplified by selection, even if the cumulative outcome was a profound reduction in regenerative ambition.

One can even see the bargain in how modern medicine must treat wounds. Surgeons clean, close, and stabilize injuries quickly; they prioritize preventing infection and restoring mechanical integrity. Reconstructive surgery can later improve form and function, but it often depends on grafts, prosthetics, and carefully staged interventions—human ingenuity compensating for biological conservatism. When medicine tries to coax regeneration—using growth factors, stem cells, engineered scaffolds—it must do so within the constraints of an immune system primed for defense and a wound-healing program primed for speed.

That is the evolutionary logic behind why a mammal meets injury with scar tissue: a rapid patch that closes the world out, keeps blood in, and reduces the vulnerable window in which a wound can become a death sentence. The scar is not only collagen. It is a signature of priorities.

When a scar forms, the body is not admitting defeat. It is making a calculation: survive the breach now, even if the repair is permanent evidence that survival mattered more than restoration.

Chapter 7

Reading the Invisible Script

For centuries, biologists watched regeneration occur like a magic trick, visible only in its final effects. Today, we have cracked the molecular seal, using high-throughput sequencing and gene editing to read the precise genetic instructions that orchestrate reconstruction-and to pinpoint exactly where the mammalian script goes silent.

7.1 The Rosetta Stone of Regrowth

A tempting idea sits at the back of almost every conversation about regeneration: perhaps salamanders and planarian worms possess a set of special, almost mythical "regeneration genes" that mammals never had. If only we could find them, the thinking goes, we could borrow them. For a long time, biology encouraged this hope. In the twentieth century, when we did not yet have genomes, it was natural to imagine that dramatic differences in ability must come from dramatic differences in parts. A lizard can regrow a tail; a child cannot regrow a finger. Different creature, different hardware.

Comparative genomics has been quietly disassembling that intuition. When the first broad lists of "wound genes" and "growth genes" were compared across animals, the surprise was not what regenerators had, but what they *did not uniquely own*. Much of the molecular cast of characters in a regenerating limb—growth factors, signaling pathways, transcription factors (proteins that bind DNA and turn genes on or off)—also lives in us. The genes that help an axolotl rebuild bone, nerve, and muscle are not alien inventions. They are variations on an ancient toolkit used to build embryos in the first place, and later pressed into service for maintenance and repair.

So where does the difference hide? If humans and axolotls share so many genes, why does one rebuild a limb while the other lays down a scar?

The most fruitful answer is not a missing ingredient but a missing *permission*. The barrier to regeneration often looks less like absence and more like silence: the right genes are present, but the network that coordinates them fails to switch on, or switches on briefly and then is shut down. In other words, the story shifts from genes as *objects* to genes as *instructions*.

To make that distinction feel concrete, it helps to use an analogy that engineers take for granted. Two machines can have similar components and still behave very differently depending on the program that runs them. Hardware matters; software decides what the hardware actually does. Genomes, too, contain both: the protein-coding sequences that build molecular tools, and the regulatory sequences that tell those tools when to act, where to act, and with whom to cooperate.

These regulatory sequences have many names. Some are *promoters*, sitting near a gene like a start button. Some are *enhancers*, which can be far away in the genome yet still influence a gene by looping through three-dimensional space. Some are *silencers* that do the opposite. Collectively, they form a control architecture: an invisible script that can be read by the cell, and, under the right conditions, rewritten by experience. Regeneration increasingly looks like a case where the script exists but is locked behind glass.

A key part of that lock is epigenetic marking, a term that sounds more mystical than it is. "Epigenetic" does not mean beyond genetics in a supernatural sense; it means beyond the DNA letters themselves. Cells package DNA around proteins and decorate that package with chemical tags. One such tag is *DNA methylation*, the addition of small chemical groups that often make a region less accessible to the machinery that reads genes. Another is a family of modifications on histone proteins that can loosen or tighten DNA packaging. These marks are how a liver cell stays a liver cell and does not spontaneously become a neuron. They are also, increasingly, how an adult mammal stays an adult mammal and does not spontaneously reopen embryonic construction projects.

The regulatory hypothesis of regeneration can be put plainly: humans do not fail to regenerate because we lack the genes to do it; we fail because we cannot reliably *deploy* those genes in the necessary pattern. The difference is not the toolbox but the foreman and the blueprint.

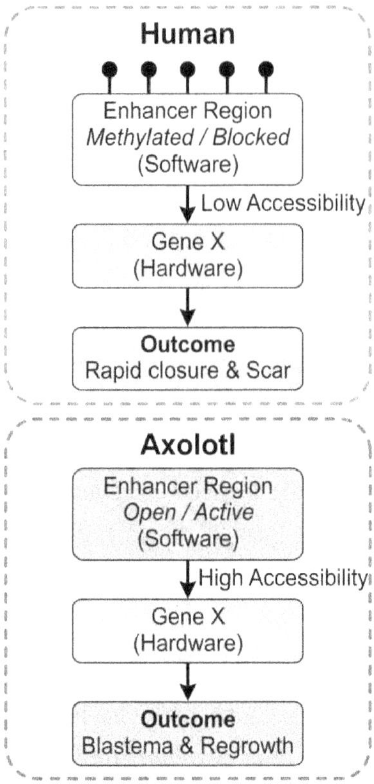

The point of such a cartoon is not to pretend that regeneration hinges on a single switch. It does not. Regeneration is a symphony, not a solo. But the diagram captures a deeper lesson that keeps resurfacing in comparative studies: what changes most between species is not the presence of a gene, but the *logic* that turns genes into a coordinated response.

That logic is easier to appreciate when one remembers what a wound asks the body to do. A cut is not merely a hole to fill. It is a crisis of identity. The tissue must decide: is the goal simply to seal the breach quickly, or to rebuild the missing structure faithfully? Speed favors a kind of biological triage: stop bleeding, block infection, patch the defect with connective tissue. Accuracy favors something else: controlled growth, careful patterning, the reopening of developmental pathways that are normally locked away after birth.

The classic anatomy of mammalian wound healing reflects the priority of speed. The immune system floods in; inflammatory signals rise; fibroblasts (connective-tissue cells) lay down collagen; blood vessels sprout; the wound contracts. The result can be lifesaving, but it is also a declaration: the body has chosen a stable patch. Scar tissue is strong, quick to produce, and relatively safe. It is also an obstacle to regrowth, because it changes the physical and chemical environment in which cells would otherwise reorganize.

Regenerating animals face the same urgent problems of bleeding and infection, yet some manage to pivot toward rebuilding. Salamanders, for example, form a structure called a blastema: a mound of proliferating cells at the stump that will later re-differentiate into bone, muscle, nerve coverings, and skin in the right arrangement. Even the word blastema carries an older style of biology, a time when one could see the mound under a microscope but could not read the molecular instructions that made it behave like an embryonic construction site.

The historical anecdote that best conveys the strangeness of regeneration comes from the era before genomes, when scientists learned by cutting and grafting and watching what happened. In the 1820s, the French zoologist René-Antoine Réaumur and others popularized careful observations of limb regrowth in crustaceans, while later nineteenth-century experimenters pushed regeneration into the laboratory. By the mid-twentieth century, researchers could do something both crude and brilliant: transplant tissue from one part of a limb to another and ask whether the regrowing structure would follow the donor's identity or the host's. These grafting experiments, refined by investigators such as Hans Spemann's intellectual descendants in amphibian development, helped crystallize the notion of *positional information*: cells behave as if they carry a memory of where they are along an axis—near the shoulder versus near the wrist, dorsal versus ventral—and regeneration succeeds when that memory is correctly updated and reconciled after injury.

The lesson from that period is easy to miss: animals do not merely regrow *tissue*. They regrow *pattern*. A salamander does not produce a generic rod of cartilage; it produces a humerus, then a radius and ulna, then carpals and digits, in appropriate proportion. Something must instruct the cells not just to multiply but to organize, to stop at the right time, to link with nerves and blood vessels, to carve joints and tendons. The very existence of such pattern suggests a deep regulatory program, more like development than like simple repair.

Comparative genomics adds a modern layer to this older story. If the same core developmental pathways are found across animals, then the decisive factor becomes how injury reconnects to those pathways. Here, non-coding DNA—the stretches that do not encode proteins—moves from the margins to the center. For decades, non-coding DNA was lazily called junk, as if it were an evolutionary attic. Regeneration research is one of the fields where that attitude has become untenable. Enhancers and other regulatory elements are often non-coding, and they can act like injury-responsive switches: silent during ordinary life, active when tissue is damaged.

A concrete example comes from work in the 2010s that identified regeneration-responsive enhancer elements in regenerating vertebrate tissues. The broad idea was straightforward: map which regions of the genome become more accessible after injury (a sign that regulatory machinery is binding there), then test whether those regions can drive gene expression in an injury-dependent way. What made the findings so compelling was the modularity: certain enhancer sequences behaved like smart sensors, turning on in specific injured tissues and staying quiet elsewhere. In effect, the genome contained a set of dormant emergency circuits that could be selectively awakened.

That modularity also helps explain why the "same gene" story can still lead to different outcomes. Suppose humans and axolotls both carry a gene involved in cell-cycle reentry—the process by which cells that normally sit quietly can start dividing again. In humans, the enhancer that would drive that gene after injury might be weak, tightly methylated, or wired to inflammatory signals that instead push cells toward fibrosis (the laying down of connective tissue). In axolotls, an enhancer near the same gene might be poised like a spring: held in check during normal life but ready to open when injury signals arrive. The protein-coding part of the gene—the hardware—may be similar; the regulatory circuitry that deploys it can be profoundly different.

This is also where the question of trade-offs becomes impossible to ignore. Mammals are not simply worse salamanders. We are different kinds of organisms with different constraints. An adult human body contains trillions of cells, many of which live for decades. A program that allows mature cells to loosen their identity, proliferate, and re-pattern tissue resembles, uncomfortably, a program that cancer tries to hijack. Regeneration requires controlled growth; cancer is uncontrolled growth. The boundary between them is policed by tumor suppressor mechanisms that mammals have refined under strong evolutionary pressure. If a salamander can reopen embryonic-like growth programs after injury without frequently tumoring, it suggests that salamanders have their own safety systems—but not necessarily the same ones, and not necessarily compatible with mammalian physiology.

In this light, the "silenced network" in mammals begins to look less like a tragic loss and more like a deliberate bargain. Fast scarring can reduce infection risk. Tight control of cell identity can reduce malignancy. Robust inflammation can clear pathogens quickly in animals that live in dirty environments. Each of these advantages has a cost: a body that is reluctant to revisit developmental plasticity may heal reliably, but it will not easily rebuild complex structures.

Comparative genomics helps identify where the bargain is enforced. One place is in regulatory regions that respond to injury signals—the molecular equivalents of smoke detectors. Injury is sensed through many channels: mechanical stress, damaged extracellular matrix, reactive oxygen species, inflammatory cytokines. These signals flow into the nucleus and converge on enhancers. If those enhancers are accessible and properly connected, they can switch on a regeneration program. If they are methylated, physically inaccessible, or wired into a different set of transcription factors, the same upstream injury signals will lead elsewhere: toward closure and scar.

Another place is in the timing of the response. Regeneration often seems to require a narrow window in which cells hover in a transitional state: plastic enough to proliferate and migrate, but still disciplined enough to remember who they are supposed to become. Development uses such windows constantly. Adult mammalian tissues tend to slam them shut quickly. A comparative lens suggests that regenerators keep more windows ajar, or can reopen them with fewer consequences.

The most emotionally charged implication of this work is also the most practical: if the genes are not missing, then the dream shifts from importing exotic parts to reactivating dormant circuits. It becomes imaginable that humans carry, in some guarded form, a version of the same script—not as a complete and readily accessible program, but as fragments, motifs, and pathways that could be recombined into something more regenerative.

One can see this in the way certain mammalian tissues behave when given special circumstances. The liver's remarkable capacity to regrow mass after partial removal is not limb regeneration, but it is a reminder that mammals are not incapable of large-scale reconstruction. Even in skin, the difference between a scar and a near-perfect repair is not always binary: fetal mammalian wounds can heal with far less scarring than adult wounds, implying that developmental context changes the regulatory landscape. The machinery for a different kind of healing exists; adulthood changes its default settings.

Comparative genomics, at its best, does not simply hand us a list of differences. It gives us a way to interpret them. When a region of non-coding DNA is conserved across species, it hints at a shared regulatory function. When it diverges or becomes epigenetically locked, it hints at a rewiring. And when a gene that seems essential for regeneration is present in humans but silent after injury, it suggests a specific hypothesis that can be tested: what switch failed to flip, and what would happen if it did?

The Rosetta Stone metaphor is apt because it captures a particular kind of translation. The original Rosetta Stone did not reveal new words; it revealed how to map between scripts. Comparative genomics does something similar. It tells us that the language of building bodies is ancient and widely shared. The mystery is why, in some animals, injury is read as an instruction to rebuild, while in others it is read as an instruction to seal and stabilize. The difference lies less in the alphabet than in the grammar: the regulatory networks that decide what counts as an appropriate response, and how much risk the body is willing to tolerate in pursuit of wholeness.

For the first time, that grammar is becoming legible. Not because we have found a single magic gene, but because we can now see the switches, the locks, and the circuits that decide whether the same genetic hardware will be used for a quick patch or for a faithful reconstruction.

7.2 A High-Definition Cell Census

A wound is a crowded place. Even a small cut recruits immune cells from the bloodstream, wakes up sleepy residents of the tissue, and alters the behavior of cells that were minding their own business minutes before. Some cells arrive only briefly, like paramedics; others settle in and start remodeling the site. Signals pass from cell to cell as proteins and small molecules, but also as physical forces: tension across the extracellular matrix, pressure from swelling, the sheer absence of neighbors where tissue has been lost.

For decades, biologists tried to understand this crowd by grinding tissue into a paste and measuring what was inside. The results were honest, and often useful, but they were averages. Imagine trying to learn how a city functions by blending all its inhabitants into a smoothie and measuring the amount of caffeine, salt, and fabric dye. You would infer that the city contains coffee drinkers, people who sweat, and people who wear blue jeans, but you would not know who runs the hospital, who drives the buses, and who is quietly keeping the water supply from collapsing. Wound sites are like that city, and regeneration is a civic project: it needs coordination among specialists, not merely increased activity in the aggregate.

Single-cell sequencing changed the scale of what can be seen. The basic idea is disarmingly simple. Instead of measuring RNA (the working copies of genes) from a whole chunk of tissue at once, one measures RNA from thousands or tens of thousands of individual cells, one cell at a time. RNA is a cell's running commentary on what it is doing: which genes are active, which programs are being called up, which identity it is holding or abandoning. Once each cell's RNA profile is captured, computational tools group cells by similarity. Those groups often correspond to recognizable cell types—macrophages, fibroblasts, endothelial cells that line blood vessels—but they can also reveal cell states that exist only briefly, in response to injury, and vanish once the job is done.

The word *census* matters. A census is not just a catalogue; it is a way of discovering who is present, in what numbers, and how their demographics change over time. In regeneration biology, this is revolutionary because the most important actors may be rare, fleeting, and easily drowned out by the loudest voices in bulk tissue. If one cell state makes up one percent of the wound, its RNA signature is reduced to a faint whisper in a blended sample. If that one percent contains the cells that actually rebuild the missing structure, losing it is like missing the entire plot because the supporting cast spoke louder.

This is why single-cell studies have been so clarifying for a concept that once felt almost mystical: the blastema. Under the microscope, the blastema is a mound of cells that appears at the end of a regenerating limb. Classic descriptions treated it as a kind of uniform mass, sometimes imagined as a pool of undifferentiated building blocks. Yet the blastema has always been suspiciously competent. It does not merely grow; it knows where it is along the limb, and it rebuilds the correct sequence of structures. That competence suggests organization hidden within apparent uniformity.

Single-cell sequencing makes that organization visible. When researchers profiled tens of thousands of cells from regenerating axolotl limbs in the late 2010s, what emerged was not a single blob but a landscape: distinct clusters of epidermal cells, immune cells, vascular cells, nerve-associated cells, and a range of mesenchymal (connective-tissue) populations that change character as regeneration proceeds. Some clusters look like familiar adult cell types; others look like transitional states, partway between a mature identity and a more plastic, proliferative one. The blastema, in this view, is not a homogeneous ball. It is a temporary neighborhood where multiple lineages gather, communicate, and, crucially, where certain connective-tissue cells adopt a progenitor-like state that can contribute to rebuilding.

That last point rewrites an older debate in a more precise language. For much of the twentieth century, the question of regeneration was framed as an argument about *dedifferentiation*: do mature cells revert to a more embryonic state? Or do resident stem cells do most of the work? Under a microscope, both possibilities can look similar. Cells in a blastema are small, densely packed, and dividing. They do not wear obvious badges announcing their origins.

Single-cell data does not settle every lineage question on its own, but it makes the questions sharper. If a cell expresses genes typical of cartilage, then begins expressing genes typical of a proliferative progenitor state, that suggests some loosening of identity. If, instead, the dividing cells express markers typical of a known adult stem-cell compartment, that supports another model. The truth can also be mixed: different tissues may use different strategies within the same regenerating limb. Muscle might behave differently from connective tissue; epidermis might supply signals rather than bulk material.

There is a historical echo here. Early embryologists were obsessed with fate maps—diagrams of which part of an embryo becomes which organ. They used dyes, grafts, and painstaking observation to trace who becomes what. Regeneration biology inherits the same obsession, but the wound makes it harder because identities are not stable; cells are in flux. Single-cell sequencing is not a dye, but it is a kind of molecular fate map, capturing snapshots of identity as it shifts.

The computational step that often draws attention—plots where each cell is a dot and dots form islands of color—can look like a modern art poster. Underneath the aesthetics lies a simple promise: if you can measure enough cells, you can see the rare islands that were previously invisible. This is especially important for regeneration because the most interesting cells may exist only for a day or two. They might also be spatially localized: present at the wound margin but absent deeper in tissue, or concentrated near nerves.

Nerves are a good example of why averages can mislead. A classic finding in salamander limb regeneration is that regeneration is *nerve-dependent*: without adequate nerve supply, the limb fails to regrow properly. For decades, this was known at the level of anatomy and surgery. Researchers could denervate a limb and watch regeneration stall. What was harder to see was the chain of molecular events: which cells respond first to nerve signals, which genes turn on, which cell states appear only when nerves are present. Single-cell studies can approach this by comparing the cellular census in normal and denervated conditions, revealing which populations fail to appear or fail to mature. The nerve, in effect, is not a mysterious life force; it is a supplier of specific signals that change the composition and behavior of the wound city.

The same logic applies to the immune system, which has been recast from a mere clean-up crew to a decision-making partner. In mammals, certain immune trajectories are associated with fibrosis and scarring. In regenerating animals, immune cells still arrive, but their timing and the balance of inflammatory versus pro-resolving states can differ. A bulk measurement might say "inflammation genes increased." A single-cell census can distinguish between different macrophage states, different subsets of T cells, and the signals they emit. It can show, for example, whether macrophages adopt a state that encourages extracellular matrix remodeling without tipping into a scar-forming program, or whether they deliver cues that help progenitors proliferate and then calm down.

Single-cell sequencing also corrects a subtler misconception: that a tissue is composed of a few fixed cell types, each with a stable identity. In many regenerative contexts, cells travel through intermediate states that are neither one thing nor another. These are not errors in measurement; they are biological realities. A fibroblast can become activated and begin secreting matrix. It can then shift into a proliferative state. Under certain conditions, it may adopt a program that resembles development more than adult maintenance. These shifts can be gradual, and single-cell data can be used to infer trajectories through what is sometimes called pseudotime: an ordering of cells that suggests a path from one state to another. It is not literal timekeeping, but it is often consistent with known stages of regeneration and can generate testable hypotheses about sequence and causality.

Planarians, the flatworms famous for whole-body regeneration, offer a particularly crisp illustration of why cell census matters. Planarians have a population of adult stem cells called neoblasts that are necessary for regeneration. Older experiments showed that if planarians are irradiated to destroy dividing cells, regeneration fails; if a single neoblast is transplanted into such an animal, regeneration can be rescued. Those results were astonishing but still left a question: are all neoblasts equivalent, or are there subtypes with different capacities?

Single-cell atlases of planarian cells in the 2010s provided an answer more nuanced than either extreme. Neoblasts are not a uniform pool. They include multiple transcriptional states, some more general, others already biased toward particular lineages. In 2018, work by Zeng and colleagues used single-cell approaches to identify a subset marked by a tetraspanin cell-surface protein and then linked that molecular identity to function by transplantation assays, demonstrating that a prospectively isolated cell could rescue a lethally irradiated planarian. This is the census delivering more than a headcount:

it connects a dot on a plot to a living capacity, turning an abstract cluster into an organism-level fact.

There is something emotionally bracing about that connection. Regeneration can feel like magic until you remember that it is enacted by real cells with particular behaviors. A single cell, correctly identified, can carry the potential to rebuild a body. That is not mystical; it is the natural consequence of an organism that keeps pluripotency (the ability to produce many cell types) in adulthood. It also highlights why mammals might be different. Adult humans generally do not keep such broadly potent cells on standby. We rely on more restricted stem cells, carefully controlled, and we use scarring as an emergency stabilization strategy.

The axolotl and the zebrafish add a different flavor of insight: regeneration in complex organs that humans also have. Zebrafish hearts can regenerate after injury in a way that mammalian hearts generally cannot. A single-cell and spatial atlas of zebrafish heart regeneration, expanded with increasingly dense sampling through time, creates something like a film rather than a snapshot. One sees cardiomyocytes (heart muscle cells) shifting into a state that allows them to re-enter the cell cycle and later re-differentiate, while other supporting populations remodel blood vessels, manage inflammation, and lay down temporary scaffolding. Without single-cell resolution, these coordinated state changes can blur into the unhelpful statement that "many genes change after injury." With it, the question becomes sharper: which cells change first, which ones supply the signals, and where does the program diverge from the mammalian response that tends toward permanent scar tissue and diminished pumping function?

Spatial information matters because a census without addresses can still be misleading. A city's demographics are not enough; where people live and where they work matters. The wound site is organized by gradients of oxygen, by proximity to nerves and blood vessels, by mechanical forces at the edge of the injury. Single-cell RNA sequencing, on its own, often loses spatial context because cells are dissociated from the tissue. The field has responded by pairing single-cell data with spatial transcriptomics and related methods that preserve location. The result is a map where molecular identity and physical position reinforce each other: one can see, for instance, a ring of wound epidermis with a distinct signaling profile, or a zone of fibroblast-like progenitors forming beneath it, or immune cells concentrated along specific routes of entry.

This marriage of census and cartography has also helped resolve arguments that once seemed intractable. During vertebrate limb development, a structure called the apical ectodermal ridge (AER) forms at the limb tip and provides signals that help pattern outgrowth. For years, there was controversy about whether axolotls recreate an AER during regeneration, or whether they use a different strategy. Cross-species atlases in the 2020s suggested a compromise that is deeply instructive: axolotls can have AER-like cells during development, but during regeneration the exact same cell type may not be reassembled as a discrete structure. Instead, AER-like signaling machinery can be distributed across multiple cell populations. This is a reminder that regeneration is not always a perfect replay of development. It can be a *re-deployment* of developmental tools in a different cellular arrangement, as if the same orchestra plays the same melody but with instruments swapping parts.

That idea fits neatly with the earlier emphasis on regulatory logic. If the key differences between species lie in gene regulation, then the same signaling program might be executed by different cells in different contexts, depending on which enhancers are accessible and which transcription factors are available. Single-cell data reveals not only *what* genes are active, but *in which cells* they are active, and that distinction can explain why an organ regenerates without needing any new genes at all.

There is also a practical side to this high-definition view. Regenerative medicine often dreams of delivering a factor—a protein or a gene—to a wound and watching tissue regrow. The single-cell census warns that such interventions are likely to fail unless they are targeted to the right cells at the right time. If a rare progenitor-like state appears only between day two and day four after injury, delivering a cue on day ten might do nothing. If a key signal must be sent from wound epidermis to underlying mesenchyme, dumping it into the bloodstream might dilute it into irrelevance. The census does not merely describe complexity; it narrows down where leverage might be possible.

It also changes how one thinks about failure. In mammals, the story is often told as "regeneration is absent." The census suggests a different phrasing: mammals often initiate parts of a regenerative response but abort it, divert it, or fence it off. Some of the same early genes turn on briefly after injury in mice and humans; some of the same pathways are activated, but the cellular cast and the timing differ. Fibroblasts may be recruited into a scar-forming trajectory; immune cells may linger in a pro-inflammatory state; epithelial closure may occur too quickly, sealing off signals that would otherwise sustain patterning. Seen through single-cell resolution, the difference between regrowth and scarring can look like a branching decision made by cells in the first days after injury.

If the older era of regeneration biology taught that pattern matters, the single-cell era teaches that pattern is negotiated cell by cell. The blastema is not a substance but a conversation. Some cells speak in growth factors, others in extracellular matrix, others in immune cytokines. Some whisper and disappear. Some become the enduring rebuilt tissue. The miracle is not that the conversation exists, but that in some animals it reliably reaches agreement.

A high-definition census does not end the mystery of why humans cannot regrow a limb, but it turns the mystery from a fog into a set of faces. One can point to a cluster of cells that appears only in successful regeneration and ask what keeps it from appearing in a mammal. One can identify the exact macrophage state associated with scarless repair in a regenerating animal and ask whether human macrophages can be nudged toward something similar. One can trace how a fibroblast-like cell moves through intermediate states and ask where mammalian cells take a different exit.

At the wound edge, where the body is forced to decide between speed and fidelity, the deciding votes are cast by individual cells. Single-cell sequencing makes those votes visible, and in doing so it replaces a vague longing for "more regeneration" with a more intimate and more demanding question: which cells must change their minds, and what would it take to persuade them?

7.3 From Observation to Authorship

A regenerating limb is a spectacle of competence, but spectacle is not explanation. Watching an axolotl rebuild a hand tells you that the body has instructions for rebuilding a hand; it does not tell you what those instructions are, nor which lines matter. Even the most detailed cell census can leave a lingering frustration: a thousand genes surge after amputation, a dozen cell states flicker into existence, and an atlas of molecular activity blooms like a weather map. The temptation is to point to the brightest storm and declare it the cause.

Biology has a long history of mistaking the loudest signal for the most important one. In the early days of molecular genetics, it was common to identify a gene that changed during a process and assume it was driving the process. Sometimes that was true; often it was a bystander reacting to the same upstream events. Regeneration multiplies this problem because everything changes at once: inflammation, blood flow, nerve activity, extracellular matrix, cell identity, cell division. Correlation is unavoidable. Causality is what we actually want.

In classical regeneration research, causality was pursued with scalpels and grafts. Remove the nerve, and regeneration fails: nerves matter. Rotate a piece of skin, and the pattern of scales changes: positional information matters. Those experiments were elegant in their own way, but they were blunt. They could tell you that a tissue is necessary, not which genes inside that tissue are necessary. They could reveal the existence of a signal, not its molecular grammar.

CRISPR changed the story by giving regeneration researchers a new kind of tool: not the ability to measure, but the ability to intervene precisely. CRISPR—Cas systems originated as bacterial immune defenses against viruses, a molecular pair of scissors guided by a piece of RNA that matches a target DNA sequence. In the laboratory, that guidance can be programmed. Provide Cas9 (the scissors) and a guide RNA (the address label), and the system cuts a chosen spot in the genome. The cell then repairs the cut. That repair can be error-prone, introducing small insertions or deletions that disrupt the gene, effectively knocking it out. With additional templates, the repair can also be coaxed to insert new sequences, allowing knock-ins and tagging.

It is hard to overstate what this means for a field that, until recently, was dominated by animals that do not fit the neat genetic workflows of mice and fruit flies. Salamanders regenerate beautifully, but they have enormous genomes, long generation times, and husbandry that demands patience. For years, researchers relied on descriptive studies and on techniques like pharmacology (adding inhibitors) or RNA interference in certain systems, each with limitations. CRISPR offered a route to make specific, targeted changes even in organisms that were not classical genetic models.

The crucial shift is from reading to writing. Once you can disrupt a candidate gene, you can ask a question that is both simple and unforgiving: if the gene is essential for regeneration, does regeneration fail when it is broken? If the gene is merely associated, does regeneration proceed anyway? If an enhancer is suspected to be a switch, can that switch be removed, muted, or rewired? The experiment becomes a form of authorship: editing the script and seeing whether the story still makes sense.

A concrete example of how authorship clarifies the plot comes from lineage tracing, a deceptively humble technique that has been transformed by genome editing. Lineage tracing asks where cells come from and what they become. Historically, it could be done with dyes, grafts, or radioactive labeling, each offering limited resolution and often fading over time. With CRISPR, it becomes possible to introduce heritable genetic marks—barcodes—that are copied as cells divide. In axolotl limb regeneration, CRISPR-enabled lineage labeling in the 2010s allowed researchers to track multiple low-frequency lineages through regeneration and compare the contribution of marked cells before and after regrowth. One of the striking findings from this work, associated with Grant Parker Flowers and colleagues (2017), was the high fidelity of regeneration: many tissues contribute in a remarkably consistent way, even across repeated amputations. Regeneration is not an improvisation; it is robust enough to reproduce a pattern of lineage contributions again and again.

That matters because it narrows the space of plausible explanations. If regeneration were simply a chaotic wound response that occasionally produces a limb-like structure, it would be unpredictable at the cellular level. The fidelity revealed by lineage tracing suggests the opposite: there are rules. A stable program constrains what cells can do and what they cannot. Once those constraints are measured, one can begin to ask which genes enforce them.

CRISPR also makes it possible to do something that used to be almost philosophical: separate the roles of a gene in development from its roles in regeneration. Many genes that are active during regeneration are also active in embryonic limb formation. If a gene is knocked out from the beginning of life, the animal may never form a limb properly, making it impossible to interpret regeneration defects later. Genome editing can circumvent this by allowing conditional strategies: editing only in a particular tissue, or at a particular time, or tagging proteins to turn them on and off with drugs or light. Even when such sophistication is not available, mosaic editing—where some cells are edited and others are not—can yield insight by allowing edited and unedited cells to compete within the same regenerating limb.

The most subtle power of CRISPR, though, lies not in breaking genes but in interrogating the non-coding switches that earlier work implicated. If the barrier to regeneration is often regulatory, then the decisive experiments must be regulatory too. Genes are hardware. Enhancers, promoters, and chromatin states are software. Editing software is how you test the idea that mammals fail not because they lack parts but because their switches are locked.

One influential line of work in the 2010s identified injury-responsive enhancer elements, sequences of DNA that become active specifically after damage in regenerating tissues. These enhancers behave like sensors: quiet during normal life, active during repair. In zebrafish and other regenerating models, such elements can drive expression of genes near injury sites, and their activity can be modular, responding in specific tissues. Kang and colleagues (2016) explored the logic of these regeneration enhancer elements and showed that they can be used to control gene expression in injury contexts, pointing to the possibility of engineering the response rather than merely observing it.

Here, CRISPR turns suspicion into proof. If an enhancer is thought to be necessary for a regeneration gene to turn on, delete the enhancer and see whether the gene's injury response collapses. If the response collapses and regeneration suffers, the enhancer was not decorative. It was part of the control circuit. Conversely, if deleting the enhancer does little, then the system may have redundancy: other enhancers compensate, suggesting a network with multiple backups. Regeneration, like any critical process, is unlikely to rely on a single fragile switch.

The same logic can be used in the other direction: instead of deleting an enhancer, insert one. This is where authorship begins to look like engineering. If an injury-responsive enhancer can be linked to an effector gene—a gene that changes cell behavior—one can build a genetic device: injury triggers enhancer activity, enhancer drives effector expression, effector alters the course of healing. In principle, this could redirect a scar-prone response toward a more regenerative one, at least in localized contexts. It is also where safety and ethics press in, because any attempt to push mammalian tissues toward greater plasticity must reckon with the risk of uncontrolled growth.

Regeneration biologists are keenly aware of this risk, often more than the public realizes. The fear is not abstract. Cancer is, among other things, a tragic parody of regeneration: cells proliferate, invade, and ignore the normal signals that tell them to stop. A successful regenerative program must do what cancer cannot: grow and then stop; rebuild and then settle back into stable, differentiated identities; invite immune participation without being destroyed by it. Any attempt to borrow regeneration's tricks must also borrow its restraint.

CRISPR is often portrayed as a magic wand, but in practice it is more like learning to type in the language a cell understands. The grammar is strict. Delivering the components into the right cells is difficult. Editing efficiency varies. Off-target effects must be monitored. In axolotls, the technical barriers were long considered prohibitive, which is why the development of practical protocols mattered so much. Work by Ji-Feng Fei and colleagues (2018), among others, helped formalize CRISPR—Cas9 genome engineering in axolotls, turning what had been an artisanal, uncertain process into something closer to a reproducible method. That kind of infrastructure rarely makes headlines, but it is the difference between a handful of heroic demonstrations and a field that can systematically test hypotheses.

Once such infrastructure exists, the experimental imagination expands. Single-cell atlases identify candidate genes and candidate cell states. Chromatin assays such as ATAC-seq identify regions of DNA that open and close during regeneration, hinting at which enhancers are active. CRISPR then allows the entire chain to be tested: perturb a transcription factor predicted to open chromatin at a key enhancer; delete the enhancer itself; tag a cell population and track its fate; alter an immune signal suspected to bias healing toward fibrosis. Each perturbation is a sentence rewritten in the script, and the consequence is read not only at the level of gene expression but at the level that matters: does tissue regenerate, does pattern return, does function recover?

A useful historical contrast makes this feel less like science fiction and more like the natural next step in a long story. In the 1860s, Gregor Mendel could not see genes; he inferred their existence through breeding patterns in pea plants. A century later, molecular biologists could see genes as sequences and could mutate them, but only in organisms with convenient genetics. Regeneration biology, for much of its history, resembled Mendel's predicament: phenomena were clear, mechanisms inferred. CRISPR is the moment when inference becomes intervention in organisms that were once genetically opaque.

There is also a quieter, more personal reason this shift matters. Many people encounter regeneration not as a laboratory curiosity but as a wound that did not heal well: a burn scar that tightens and aches, a heart attack that leaves a permanent deficit, a spinal cord injury that does not resolve. For these conditions, correlation has never been enough. Patients have been offered associations and risk factors, not levers. The promise of authorship is the promise of levers: to identify not just what happens in damaged tissue, but what must be changed for a different outcome.

Even in its most immediate form, authorship can be surprisingly modest. One does not need to imagine regrowing an arm to see value. If a regulatory element can be engineered to drive anti-fibrotic factors only in injured heart muscle, reducing scar formation and improving function, that would be transformative. If macrophage behavior can be nudged at the right time to favor resolution rather than chronic inflammation, many tissues might heal better even without full regeneration. Regeneration, in this sense, is not an all-or-nothing miracle; it is a direction in which biology can be pushed.

The deeper lesson is methodological. Reading the invisible script is not enough, because every complex process produces a symphony of signals that can be mistaken for causes. Authorship is how one learns which notes are essential. When a regenerating animal fails after a precise edit, the failure is informative in a way that descriptive data cannot match. It tells you that a particular element is not merely present but necessary, not merely active but decisive.

In the most compelling experiments, the edit does not simply break regeneration; it breaks it *in a specific way*. A limb may regrow but with patterning defects, suggesting that growth occurred but positional information was lost. A wound may close but never form a proper blastema, suggesting a failure in early cellular reprogramming. An organ may rebuild tissue but fail to revascularize, suggesting that the script for blood vessels was not properly engaged. Specific failure modes are a form of understanding: they reveal the architecture of the program by showing where it can be interrupted.

A successful regeneration program is an improbable peace treaty among competing needs: speed and safety, growth and restraint, plasticity and identity. CRISPR does not guarantee that the treaty can be renegotiated in mammals, but it provides a way to test which clauses matter, and whether the silences in our own genome are locks that can be opened or safeguards that must be respected. In the hands of careful experimentalists, the technology turns the ancient question—why can't we do what salamanders do?—into something sharper and more intimate: which edits change the wound's decision, and which edits simply make the body defend itself in a different way.

Chapter 8

Awakening the Dormant Code

While salamanders regrow limbs effortlessly, humans typically settle for scars, yet our bodies harbor ghostly remnants of this superpower. We investigate the rare instances where human biology breaks its own rules-from regrowing fingertips to expanding livers-and explore how modern medicine attempts to coax our cells into remembering their embryological origins.

8.1 The Fingertip and the Liver

There are two kinds of hope that gather around the word *regeneration*. One is the everyday kind: the cut that closes, the bone that knits, the bruise that fades from purple to yellow to nothing. The other is rarer and stranger: the hope that a missing piece returns with its original architecture, as if the body kept a blueprint in a drawer and simply unfolded it again. Humans live almost entirely in the first world. We heal, but we do not usually *rebuild*. Yet two parts of the human body behave as if they still remember, faintly, an older set of instructions. One is the liver, which can restore lost mass with almost unnerving confidence. The other is the fingertip, which, under surprisingly specific conditions, can regrow what was taken away.

These two examples are often discussed together, as if they were variations on the same miracle. They are not. They are closer to two different dialects of repair, sharing a few familiar words but disagreeing on grammar. The liver is a champion of *compensatory growth*: when tissue is lost, the remaining tissue expands, restoring function without necessarily recreating the original pattern. The fingertip, in contrast, can sometimes mount something nearer to *epimorphic regeneration*: the creation of new structures through a transient, embryonic-like rebuilding phase. The difference matters, because it points to what humans still possess, and what has been shut down.

Consider how a surgeon speaks about the liver. In the operating theatre, the liver is not treated like a fragile heirloom. It is treated like a practical organ with slack in the system. Living-donor liver transplantation depends on that slack: a healthy adult can donate a portion of liver to a child or another adult, and both donor and recipient can, over time, regain near-normal liver mass. Behind the scenes is a choreography of signals—inflammatory cues, growth factors, changes in blood flow and tissue tension—that drive hepatocytes (the liver's main working cells) to re-enter the cell cycle. The liver is one of the few adult human organs in which large numbers of mature cells can be persuaded to divide without collapsing into chaos.

This has a pedigree in experimental biology as well as medicine. In 1931, Higgins and Anderson described a standardized partial hepatectomy in rats that removed about two-thirds of the liver. The technical achievement was not only that the animals survived, but that the remnant liver enlarged and restored mass in an orderly way. That simple surgical trick became a cornerstone of regeneration research for the next century. It offered something scientists love: a repeatable, quantifiable injury with a dramatic response. Remove a defined fraction; watch the organ swell back toward its old weight.

Yet the very success of the liver can mislead. If a salamander loses a limb, the rebuilt limb has joints in the right places and muscles attached along the correct lines. It does not merely become a thicker stump. The liver's achievement is different. It is less about recreating a missing *shape* and more about restoring adequate *capacity*. In many cases, the lobes that remain enlarge, rather than the missing lobe being reconstituted as a distinct anatomical structure. What returns is mass and function—not a precise, segment-by-segment reprinting of what was cut away.

There is a deep reason for this difference: a liver is, in some sense, modular. Its essential job is biochemical. It filters blood, processes toxins, manages glucose, manufactures proteins, packages bile. If you have enough healthy hepatocytes arranged with enough vasculature and ductwork, you have enough liver. A fingertip, by contrast, is an engineered object. It has to be the right *length*. It has to have a nail with a curved plate and a bed. It has to carry nerves that end in sensory structures, and skin with its specialized ridges. It has to coordinate bone, tendon, fat pad, and epidermis in a tight, layered geometry. Restoring *volume* is not enough; the fingertip must restore *pattern*.

That is why the fingertip is so scientifically provocative. When it regrows, it hints that the adult human body is not entirely ignorant of pattern formation. The ignorance is conditional.

Clinicians have known for decades that the tips of children's fingers can sometimes regrow after amputation. The stories circulate in emergency rooms and family lore: the toddler who slammed a finger in a door, the frantic ride to the hospital, the tiny severed piece wrapped in gauze, and then, months later, the finger that looks eerily close to what it used to be. But the fine print matters. This is not the regrowth of an entire finger. It is typically the distal tip, often when the amputation is relatively far out—at the level where the nail and its surrounding tissues remain partly intact. The more proximal the injury, the less likely a true regrowth becomes, and the more likely the outcome is scarring, a shortened digit, or a rounded stump.

That boundary—distal versus proximal—is a clue that the key player is not the bone alone. It is the *nail organ*. In mouse experiments, amputations that preserve the nail-forming region are far more permissive for regeneration than those that remove it. The nail is not merely decoration or protection; it is a living epithelial structure with its own stem cells and signaling dynamics. The nail matrix, tucked under the proximal nail fold, is a factory line: stem-like cells produce keratinized nail plate that slides forward as it grows. In the language of developmental biology, it is a niche, a microenvironment that sustains a particular kind of cellular identity.

After a distal digit tip amputation in mice, researchers have observed the formation of a *blastema*—a mound of proliferating cells that appears under the wound epidermis. The term can sound like science fiction, but it is simply a name for a temporary construction site: cells gather, multiply, and coordinate, then later differentiate into bone, connective tissue, and other components. In salamanders, the blastema is a classic hallmark of limb regeneration. Seeing something analogous in a mammal is like finding an old tool in the back of a drawer: dusty, seldom used, but still recognizably functional.

What turns that tool on? One of the clearest signals implicated is Wnt/β-catenin, a pathway famous for its role in embryonic development and adult stem cell maintenance. In work by Takeo, Chou and colleagues (2013), Wnt activity in nail epithelium was linked to digit regeneration in mice; disrupting the pathway impaired nail regrowth and digit tip regeneration. In a complementary vein, Lehoczky and Tabin's group (2015) identified Lgr6 as a marker of nail stem cells required for digit tip regeneration, strengthening the idea that a specific epithelial stem cell compartment is part of the regenerative engine rather than a passive bystander.

The fascination here is not that Wnt exists. Wnt signals are used all over the body. The fascination is *where* the signal originates and *what it seems to authorize*. The nail epithelium appears to do more than repair itself; it helps coordinate deeper tissues, including bone. It behaves like an organizer, a word that, in developmental biology, carries a kind of reverence. An organizer is a tissue that tells its neighbors what to become and where to go, translating a local signal into a structured outcome. In the fingertip, the nail's epithelial niche may be giving a local tissue environment permission to rebuild rather than merely close.

To appreciate how odd that is, it helps to think about what the body is normally trying to do after injury. The default priority is speed and containment. A wound is an open border. It is an invitation to infection and fluid loss. It is also a mechanical breach: tissues under tension, now exposed. The body's rapid response is inflammation, clotting, and then the recruitment of fibroblasts, the connective-tissue cells that lay down collagen. Fibroblasts are the emergency contractors of adult mammalian healing. They do not rebuild the original building; they erect a sturdy patch. The patch is a scar: strong, often stiff, and usually lacking the original architecture of skin, muscle, or organ.

In that context, regeneration is not merely hard. It is actively opposed by a program that usually keeps us alive. Every time a wound heals without infection, without bleeding out, and without leaving the tissue structurally unstable, that program wins its argument.

The liver's success can be understood as a clever compromise with this reality. It regenerates not by building a complex external shape, but by expanding what is already there. Much of its architecture—the microscopic arrangement of hepatocyte plates, bile canaliculi, and sinusoids—is repetitive. Expansion can preserve function without requiring elaborate repatterning. The liver also sits in a chemical environment that is used to managing insults: toxins, metabolic stress, fluctuating nutrient loads. It has evolved a robust capacity to proliferate and then stop, to expand and then terminate the expansion before it becomes a malignancy. That stopping is crucial. Organ growth is dangerous when it forgets how to end.

By contrast, a fingertip has to do something closer to sculpture. It has to recreate a distal phalanx contour, rebuild the nail apparatus, and restore a sensate pad. That kind of rebuilding is so rare in adult mammals that it is tempting to label it an anomaly and move on. But anomalies are where the rules become visible.

One rule is that *pattern formation seems to require a permissive epithelial niche.* The nail organ is a specialized epithelium that retains stem cells, and those stem cells retain developmental signaling activity. Another rule is that *regeneration seems to require a temporary suspension of adult rigidity.* The blastema-like state is, by design, a return to plasticity: cells proliferate, migrate, and then redifferentiate. Plasticity is power, but it is also peril. The line between a controlled blastema and a tumor is not drawn with ink; it is drawn with safeguards.

This tension becomes clearer when we place the fingertip and the liver side by side, not as two miracles, but as two solutions to the same problem: how to restore integrity without unleashing uncontrolled growth.

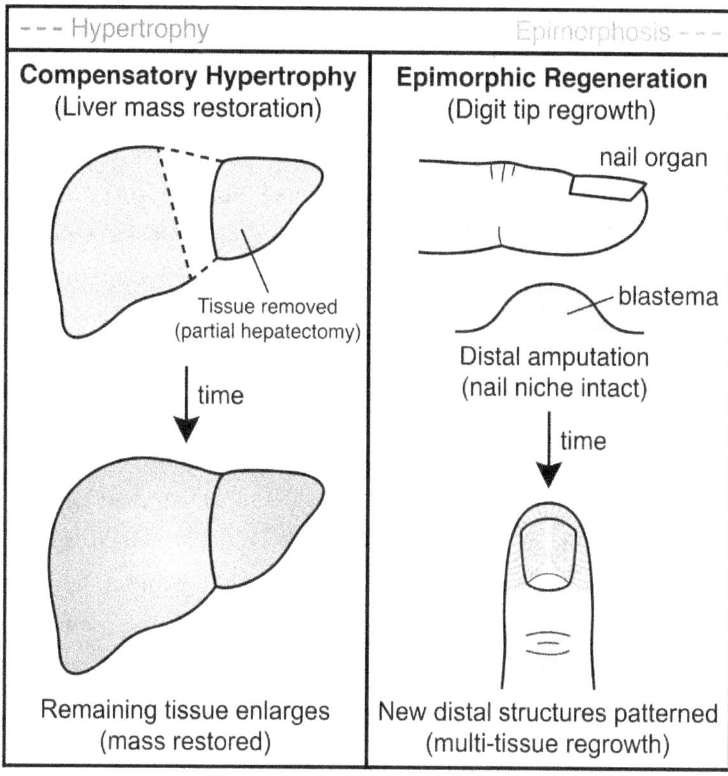

The diagram makes a blunt point: one process fills in, the other rebuilds. Both require cells to proliferate, but proliferation is not the essence of regeneration. Proliferation is a *resource*. Pattern is the *instruction*.

That distinction can be brought down to a simple observation. If the liver restored mass by forming a blastema-like mound that then reassembled a lobe in the old shape, we would recognize it as the same kind of regeneration seen in salamanders, translated into an internal organ. It does not behave that way. If the fingertip merely thickened the stump, restoring some padding but not the nail, bone contour, or tactile apparatus, we would call it robust healing. Instead, in favorable conditions, it performs a constrained but genuine redevelopment.

Even so, the fingertip's talent is not unlimited. Mouse studies suggest that regenerated bone can be imperfect—shorter than the original or altered in contour. Human regrowth, when it occurs, may leave subtle differences in nail shape or fingertip padding. This is not a perfect replay of embryogenesis; it is a pragmatic reconstruction in hostile territory, undertaken by adult tissues that are simultaneously trying to prevent infection, close the wound, and keep the limb functional.

That hostile territory includes a powerful mechanical factor: *contraction*. Adult mammalian wounds tend to contract as they heal, driven by myofibroblasts that pull on the extracellular matrix like tiny winches. Contraction narrows the wound quickly, which is useful, but it also encourages aligned collagen deposition and a stable scar. A scar is not merely different tissue; it is a different *geometry* of tissue, with collagen fibers arranged to resist future tearing rather than to support fine structure. In the fingertip, where fine structure is the whole point, too much contraction can be a kind of veto. It closes the border, but it closes it with a material that refuses to be remodeled into nail bed, dermal ridges, and organized bone.

Here the age effect begins to make intuitive sense. Children's tissues, on average, are more plastic. Their immune responses differ, their fibroblasts behave differently, and their epithelial niches are in a state of active growth. A child's fingertip is not an adult fingertip scaled down; it is a developing structure with developmental programs still warm. That does not guarantee regeneration, but it lowers the threshold. It is easier to coax a system already tuned for growth into one more bout of organized rebuilding than to force an adult system to reverse its default settings.

The liver again shows the other side of the coin. It is adult, it is fully differentiated, and yet it repeatedly permits large-scale cell cycling. Why does that not produce cancer every time? Part of the answer is that the liver's proliferation is tightly staged: a priming phase, a proliferative phase, and a termination phase. The details are molecular and intricate, but the design principle is not: growth comes with a built-in off switch. The Hippo pathway and its effectors YAP/TAZ, for example, have emerged as central regulators of organ size and regenerative proliferation in the liver; they can promote regrowth when needed and restrain it when not. The price of such powerful levers is that they sit uncomfortably close to oncogenesis. Push too hard, for too long, and the same signals that restore tissue can also help transform it.

That price is not hypothetical. Many pathways that support developmental growth and stem cell behavior are frequently implicated in cancers. Wnt signaling is a prime example in other contexts; YAP/TAZ activation is another. The body's reluctance to regenerate complex structures may not be laziness or lack of imagination. It may be a defensive posture, honed over millions of years: do not re-open embryonic plasticity in adult tissues unless the benefit is overwhelming and the system can guarantee a safe landing.

The fingertip suggests that one safe landing zone still exists. It is small, distal, and bounded by a specialized epithelial organ that can coordinate the response. The liver suggests another: an internal organ with repetitive micro-architecture, whose main demand is mass and metabolic throughput, and whose growth is normally constrained by powerful homeostatic circuits.

Taken together, these examples do not say that humans are on the cusp of regrowing limbs. They say something both more modest and more useful: the genetic and cellular machinery for significant rebuilding has not vanished. It has been fenced off behind conditions. The fence posts look like niches (such as the nail organ), mechanics (such as contraction versus open remodeling), immune tone (inflammation as either support or sabotage), and growth control (pathways that must accelerate and then brake).

If there is a single emotional lesson in this, it is that the body is not a failed salamander. It is a body that has chosen reliability over exuberance. Most days, that choice is invisible. It becomes visible only when a child's fingertip quietly returns, nail and all, or when an adult's liver grows back the mass that a surgeon removed. In those rare moments, the body reveals that it still knows more than it usually allows itself to say.

8.2 Architects of Flesh

A wound is a negotiation carried out at speed. The skin breaks, blood clots, immune cells arrive, and within hours the body begins choosing between two futures. One future is quick, strong, and blunt: close the gap, lay down collagen, restore a barrier, move on. The other is slower and more ambitious: rebuild what was there, with the right shapes and interfaces, so that the repair is hard to detect. Adult mammals usually take the first option not because they lack imagination, but because the second option requires a kind of cellular looseness that can turn treacherous.

The central problem is not that cells cannot grow. They grow all the time. The problem is that adult tissues have been trained to treat open-ended growth as a threat. When an adult human heals a deep cut, the body does not ask, "How do I recreate the original skin?" It asks, "How do I stop this from getting worse?" Fibroblasts answer that question with scar.

A fibroblast is an unglamorous cell with an unglamorous job: make connective tissue. Under normal conditions, fibroblasts help maintain the extracellular matrix, the mesh of proteins and sugars that gives tissues their mechanical character. After injury, fibroblasts become more aggressive and more muscular. Some transform into myofibroblasts, cells that can pull on the matrix and contract the wound, cinching it shut. They also secrete collagen in thick bundles. Collagen is strength in a hurry. It is also the reason scars look and behave differently: the fibers are often aligned and dense, leaving tissue stiff and less functional.

Regenerative biology, in practice, is the art of interrupting that default. Much of the modern field could be summarized as an attempt to build *architectural scaffolding* for cells—not only literal scaffolds, but chemical, electrical, and mechanical contexts that coax tissues to rebuild rather than plaster over. The goal is not to abolish wound healing; it is to persuade the body to pause at the fork and take a different turn.

One way to see the fork clearly is to imagine the earliest hours after injury as a race between programs. In one lane: inflammation, clotting, fibroblast recruitment, contraction, collagen deposition, and eventual remodeling into scar. In the other: a transient cover of specialized wound epidermis, a permissive signaling environment, controlled proliferation, and pattern formation. In salamanders, the second lane is so well practiced that it looks effortless. In adult mammals, the first lane has a head start.

There are three broad strategies being pursued to change that race: rebuilding tissues outside the body and implanting them (organoids and engineered tissues), forcing adult cells to behave as if they were in a developmental context (bioelectric and signaling-based reprogramming), and reshaping the wound environment so that scarring is mechanically disfavored (decellularized matrices and anti-contraction scaffolds). These strategies often overlap; the boundaries blur in the clinic and in the lab. What unites them is a sober recognition that regeneration is not just a gene switch. It is a local state of matter.

The oldest and most concrete strategy is *replacement*: if the body won't rebuild, we build for it. This is the logic behind skin grafts, one of the earliest successful tissue interventions. During the Second World War, burn medicine advanced rapidly under the pressure of necessity. Surgeons refined techniques for harvesting split-thickness grafts and for moving skin to cover wounds. A graft is not regeneration, but it is an engineered workaround: it provides a living barrier, reduces infection risk, and can limit the degree of contraction and scarring that would otherwise occur.

Modern regenerative medicine tries to move beyond grafting sheets of skin to recreating more complex tissues. The emblematic example is the organoid, a tiny, self-organizing structure grown from stem cells that can reproduce some features of an organ. An intestinal organoid, for instance, can form crypt-like folds and differentiated cell types that resemble the lining of the gut. A brain organoid can develop layered neural regions and electrical activity, albeit in simplified form. Organoids are not miniature, functional organs. They are proof that cells carry an intrinsic capacity to arrange themselves when the environment is permissive.

The most striking thing about organoids is what they do *without* an engineer's micromanagement. Provide the right growth factors, the right gel-like matrix, and enough time, and the cells begin to sort and pattern. This is the developmental program resurfacing. In adult wound healing, the developmental program is usually muffled by inflammation and fibrosis. In a dish, with a carefully tuned chemical soup and a soft matrix, it can speak more freely.

Clinically, the hope is to use organoid-derived cells to repair tissues that scar badly: the heart after a myocardial infarction, the brain after stroke, the spinal cord after injury. Yet the dish is kind. The body is not. When cells are transplanted into a damaged adult tissue, they enter a storm of inflammatory signals, reactive oxygen species, mechanical stiffness, and immune surveillance. The storm tends to push them toward survival mode, not architectural ambition. That is why many organoid-based approaches now emphasize the *microenvironment* as much as the cells: the support matrix, the immune modulation, the vascularization. Without blood supply, transplanted cells die. Without mechanical compatibility, they differentiate oddly or trigger fibrosis. Without immune control, they are rejected or inflame the site further.

A second strategy attempts something more audacious: rather than importing a new structure, persuade the existing tissue to rebuild itself by changing its *information state*. Here, "information" is not metaphorical. Cells are constantly reading signals: chemical gradients, contacts with neighbors, stiffness of the matrix, oxygen levels, electrical potentials across membranes. These inputs are integrated into gene expression and behavior. In regenerating animals, certain information patterns reliably appear at wounds: specific wound epidermis signals, nerve-derived cues, bioelectric patterns that correlate with where structures should form.

Bioelectricity is especially interesting because it offers a way to influence cell behavior without directly editing genes. Every cell maintains a voltage difference across its membrane, created by ion channels and pumps that move sodium, potassium, chloride, and other ions. These voltages are not merely housekeeping for neurons. They are present in skin, muscle, and embryonic tissues, and they can influence proliferation, migration, and differentiation. In some experimental systems, altering ion channel activity changes how tissues pattern themselves, as if the voltage landscape were a kind of map.

The appeal of bioelectric interventions is their reversibility. Genes are hard to un-edit. A voltage pattern can, in principle, be shifted for a time and then allowed to settle. Electrical stimulation has already entered the medical mainstream in more modest forms: bone healing can be assisted by electrical or electromagnetic stimulation in certain non-union fractures; nerves and muscles are routinely stimulated in rehabilitation; deep brain stimulation treats Parkinson's symptoms. These applications do not create new organs, but they demonstrate that tissues listen to electrical cues.

In the regeneration context, one can imagine therapies that deliver carefully patterned electrical stimulation to a wound, not to make muscle contract, but to bias the behavior of skin cells, immune cells, and fibroblasts. The difficulty is that biology is not a single wire. A wound is a tangled circuit. If the aim is to push healing away from fibrosis, the stimulation would need to be tuned not only in intensity but in timing, geometry, and the tissue's changing conductivity as it swells and remodels.

The third strategy is the most physically grounded: change the wound's mechanics so that scarring becomes less likely. Scar is not only a chemical outcome; it is a mechanical one. When myofibroblasts pull, they align collagen. When collagen aligns, it provides a template that reinforces the pull. The wound stiffens in a particular direction; cells sense that stiffness and respond by producing more aligned matrix. It is a feedback loop, a kind of architectural momentum that ends in a scar that is strong but inelastic.

This is where scaffolds and decellularized matrices enter. A decellularized scaffold is tissue that has had its living cells removed, leaving behind the extracellular matrix. The idea is to strip an organ or tissue down to its structural proteins and architecture, then repopulate it with cells. In the most ambitious versions, a whole organ is decellularized, leaving a ghostly framework of vessels and chambers that can then, in theory, be reseeded with appropriate cells. Even when whole-organ engineering remains difficult, decellularized matrices are already used as biologic dressings and surgical materials.

The key concept is that the extracellular matrix is not inert stuffing. It is a signaling platform. Cells attach to it through receptors called integrins and other adhesion molecules, and those attachments influence gene expression. If the matrix is too stiff, cells behave differently than if it is soft. If the fibers are aligned, cells migrate and deposit new matrix along those lines. If the matrix contains particular molecular motifs, immune cells respond differently. A well-designed scaffold can, in effect, whisper to cells: "Build this way."

A particularly provocative line of thinking argues that adult mammals fail to regenerate largely because wounds contract too quickly and too forcefully. If contraction is reduced, the environment might remain open long enough for pattern formation to proceed. Some scaffold designs aim to do exactly that: physically interrupt the contraction machinery by providing a porous structure that myofibroblasts infiltrate but cannot align into a single pulling direction. The result can be less aligned collagen and, in some animal models, partial restoration of more normal tissue architecture rather than a dense scar. The image is almost architectural: instead of letting workers pull the walls inward and pour concrete, you give them a flexible framework that forces a slower, more organized rebuild.

These mechanical approaches have an intuitive appeal because they treat scarring as a default *engineering solution* rather than a moral failure of biology. If a bridge collapses, the first priority is a temporary support so people can cross. Only later do you rebuild the elegant structure. Adult mammalian healing rarely gets to the elegant part because the temporary support hardens into permanence. A scaffold can be designed to keep the site in a rebuildable state, delaying the "harden into permanence" step without leaving the body exposed.

The most interesting developments occur where these strategies intersect. Imagine a wound dressing that is not merely a barrier but a computational device: it measures local pH, oxygenation, inflammatory markers; it adjusts electrical stimulation; it releases growth factors in pulses; it provides a matrix that discourages contraction; it delivers cells or organoid fragments at the right time. This sounds futuristic, but pieces of it exist. Smart bandages are being prototyped. Controlled-release biomaterials are routine in drug delivery. Electrical stimulation devices are common. The challenge is integration, and integration is where biology becomes humbling.

A concrete example helps. Consider a full-thickness skin wound. If left to heal on its own, it will contract and scar. If covered with a conventional dressing, it may heal faster and with less infection, but the scar will still form. If treated with a skin substitute that includes a dermal-like matrix, the contraction can be reduced and the resulting tissue can be more pliable. Now imagine adding cells: fibroblasts tuned to produce a more normal extracellular matrix, keratinocytes to re-epithelialize, perhaps endothelial cells to encourage vascularization. Even then, scarring may persist because the immune environment is still primed for fibrosis. So the therapy may incorporate anti-inflammatory signals or molecules that bias macrophages (key immune cells at wounds) toward a pro-repair, less fibrotic behavior.

The time dimension is crucial. Early inflammation is not purely destructive; it clears debris and fights infection. Suppress it too much and healing stalls or infection blooms. Fibroblasts are not villains; without them, wounds can gape and fail. The goal is not to erase the wound-healing cascade but to change its *sequence* and *end point*. In practice, that may mean letting the body seal the wound quickly, then reopening a window of plasticity under controlled conditions, or it may mean prolonging a developmental-like state long enough for better patterning, without tipping into chronic inflammation.

A historical anecdote from a different corner of medicine illustrates how hard it is to change the body's defaults. When organ transplantation first became viable, surgeons could connect blood vessels and ducts, but the immune system treated the organ as an invader. The major breakthrough was not only surgical technique but immune modulation. The body did not need to be taught how to perfuse a kidney; it needed to be taught not to destroy it. Regenerative therapies face a parallel problem. Cells may know how to build. The wound environment must be persuaded not to shut the project down with collagen and contraction.

There is also a conceptual trap worth avoiding: the idea that scarring is merely a *lack* of regeneration. Scarring is an *active* program, with its own signals, timelines, and advantages. That is why so many anti-scar approaches fail when they simply block a single molecule. The system has redundancies. If you inhibit one pro-fibrotic pathway, another may compensate. If you blunt contraction, the wound may remain open and inflamed, generating a different kind of pathological remodeling. The architects of flesh do not fight a void; they negotiate with a stubborn, adaptive bureaucracy.

The most hopeful sign is that small successes tend to reinforce one another. A matrix that reduces contraction can also reduce the signals that recruit more myofibroblasts. A calmer immune environment can allow transplanted cells to survive long enough to influence remodeling. Electrical cues can steer migration and proliferation in ways that make the physical scaffold more effective. Regeneration, in this framing, looks less like turning on a hidden gene and more like getting a committee to agree—cells, matrix, nerves, immune system, blood supply, mechanical forces—that rebuilding is permitted.

Even in humans, glimpses of that agreement already exist at the margins: the fingertip when the nail niche remains, fetal wounds that can heal with remarkably little scarring, the liver when mass must be restored and termination signals hold firm. The ambition of current science is to reproduce the conditions of those margins in tissues that have forgotten how to cooperate.

If scarring is the body's emergency patch, then overriding it requires making the emergency feel over. That is a delicate psychological trick, except the psychology belongs to cells. They must be convinced that the world is stable enough to risk complexity: that oxygen will be adequate, that infection is controlled, that mechanical forces are not tearing the site open, that growth will be restrained once the job is done. When that conviction takes hold, even briefly, tissues can begin to behave less like a triage unit and more like a construction crew, laying down something closer to the original design rather than the fastest possible fix.

8.3 The Promethean Bargain

Regeneration has an easy public image: a lost limb returning, a spinal cord reconnecting, an old scar dissolving into healthy skin. The hard part is not imagining the benefits. The hard part is looking directly at what the body must do in order to achieve them. To regenerate in the adult, cells must become more adventurous than they are normally allowed to be. They must multiply, migrate, change identity, and cooperate across tissues that have spent decades maintaining boundaries. In a laboratory, those behaviors look like promise. In an oncology clinic, they look uncomfortably familiar.

The bargain sits in a single phrase that keeps resurfacing in regenerative biology: *plasticity*. Plasticity means the capacity of a cell to change what it is and what it can do. During embryonic development, plasticity is normal and necessary. In adult life, plasticity is useful in narrow places—the lining of the gut, the blood system, the skin—and carefully restricted elsewhere. Many of the strategies that aim to unlock regeneration do so by increasing plasticity, even if they do not use the word. They push cells to loosen their identities, to re-enter the cell cycle, to act less like settled citizens and more like builders on a rapidly changing site.

Cancer is, among other things, a disease of mismanaged plasticity and mismanaged growth. A tumor cell is a cell that has found ways to keep dividing, to ignore stop signals, to recruit blood vessels, to evade immune destruction, and often to adopt new states that help it survive. The overlap is not accidental; both regeneration and cancer draw from the same molecular tool chest. The difference is governance.

This is why the liver, so often held up as evidence that humans can regenerate, is also a warning. The liver can restore mass because adult hepatocytes can be coaxed back into proliferation. But hepatocytes are not simple, pristine units waiting to be activated. Many are polyploid: they carry extra copies of their genome, sometimes as a normal feature of maturation. Polyploidy can be protective in some contexts, a buffer against damage, a way to maintain function despite stress. Yet when a polyploid cell attempts to return to a proliferative state, the bookkeeping becomes complicated. Under certain circumstances, polyploid hepatocytes can undergo ploidy reduction as they divide, and that reduction can increase the risk of chromosomal missegregation. Chromosomal instability is a classic ingredient in tumor initiation. The very flexibility that makes liver mass restoration possible can, if misdirected or prolonged, become an entry point for malignancy.

The relevant point is not that liver regeneration *causes* cancer in any simple way. Most people's livers regenerate after partial injury without transforming into tumors. The point is that the pathways that permit regrowth are the same ones that, in different settings, are implicated in cancer. That is why growth is always paired with brakes.

One of the most studied braking systems is the Hippo pathway, a name that sounds whimsical until you learn what it governs: organ size. When Hippo signaling is active, it keeps the transcriptional coactivators YAP and TAZ out of the nucleus, limiting growth-promoting gene expression. When Hippo signaling is reduced, YAP/-TAZ enter the nucleus and drive programs associated with proliferation, survival, and sometimes a more stem-like state. In mouse genetics, removing YAP/TAZ function can impair efficient liver regeneration after partial hepatectomy; the organ struggles to recover mass. Conversely, chronic hyperactivation of YAP/TAZ has been associated with cancers in multiple tissues, and prolonged activation in experimental systems can lead to tumor formation. The biology reads like a parable: weaken the accelerator and the tissue cannot regrow; jam it down and you may not like where you arrive.

Wnt/β-catenin signaling occupies a similar double role. It helps sustain stem cell compartments and is involved in patterning during development. It is also a pathway frequently altered in cancers, particularly in tissues like the colon where stem-cell driven renewal is constant. In the fingertip, Wnt activity in the nail epithelium appears to support a regenerative response. In the wrong context, the same pathway can support uncontrolled expansion. The ethical dilemma is not about whether these signals are good or bad. It is about whether medicine can use them briefly, locally, and reversibly, without leaving behind a cellular rebellion.

A useful analogy comes from aviation rather than mythology. A fighter jet is designed with enough power to be agile, but it is also designed with control surfaces and stability systems to prevent catastrophe. If you give a commuter airplane a fighter jet's engine without the stability systems, you do not get a better commuter airplane. You get an accident. Many adult tissues are commuter airplanes: reliable, stable, optimized for long-term maintenance. Regenerative therapies propose, in effect, to install a more powerful engine of growth and plasticity. The question is whether the control systems can be engineered with equal sophistication.

Risk is not only biological. It is also social. The phrase *body shop* is sometimes used half-jokingly to describe a future clinic where tissues are repaired or replaced as routinely as car parts. The joke hides something sharp. Cars are repaired according to the owner's budget and the market's interest. Humans do not want their organs and limbs to be repaired on those terms.

The history of medicine offers a sobering guide to how unequal access can become, even for interventions that are eventually commonplace. Dialysis, when it first became available, was not simply a medical procedure; it was a rationed resource. In the early days, committees in the United States made decisions about who would receive life-saving treatment. The criteria were a blend of medical prognosis and social judgment, and the very existence of the committees became a public scandal. The situation improved as technology expanded, costs shifted, and policies changed, but the lesson remains: when a therapy is expensive and life-altering, society tends to allocate it unevenly, especially at first.

Regenerative medicine, if it succeeds, is likely to begin as expensive, specialized, and limited. A therapy that restores a damaged heart muscle, prevents fibrosis after spinal injury, or enables a complex tissue to be rebuilt rather than scarred would not merely relieve suffering. It would extend functional life. That makes it a form of longevity medicine, even if it never claims that title. The benefits will not be distributed at random.

This creates a second bargain: the bargain between individual desire and collective fairness. If a person can pay for a therapy that reverses joint degeneration, restores vision, or regrows tissue lost to trauma, it is hard to argue that they should not. Yet if the result is a society where those with resources accumulate not only wealth but also stronger, longer-lasting bodies, inequality becomes embodied. It is not only that some people live better; it is that their biology is maintained at a higher standard. The old metaphors of "haves" and "have-nots" take on a literal edge.

There is also a third bargain, quieter but more personal: the bargain between repair and identity. Scars can be disfiguring and painful, but they can also be meaningful. For some, a scar is a memory of survival. The ability to erase scars, to reset tissue to an earlier state, raises questions that are not only cosmetic. If a war injury can be regenerated away, what happens to disability identity? If a childhood burn can be undone in adulthood, how does that change a person's relationship to their own history? These are not arguments against healing. They are reminders that bodies are not just machines; they are narratives.

Ethics in regenerative medicine is often presented as a matter of consent and safety: do patients understand the risks, and have those risks been minimized? Those are necessary questions, but insufficient. The risks here are not always immediate. A therapy that briefly increases cellular plasticity might not cause problems for years. Cancer risk is not a side effect like nausea; it is a long shadow. This makes informed consent harder, because it requires discussing probabilistic futures. It also makes regulation harder, because the evidence takes time to accumulate. A treatment that looks safe in a two-year study may look less comfortable in a ten-year follow-up.

The temptation, for patients and for companies, will be to treat the regenerative switch as something that can be turned on and off like a light. Biology rarely cooperates. A growth factor pulse may initiate a cascade. A scaffold that changes mechanics may alter immune behavior for months. A transplanted cell population may persist, integrate, and evolve under selection pressures. Even gene editing approaches that target a specific locus can have pleiotropic effects—one genetic change influencing multiple traits—because the pathways involved in regeneration are deeply interwoven with metabolism, immunity, and development.

This is where the fetal comparison becomes ethically revealing. Fetal wounds can heal with minimal scarring, and that fact is often invoked as a kind of natural proof that scarless healing is possible in mammals. Yet fetal biology is not simply adult biology with a different setting. The fetus exists in a distinct immune environment, with different inflammatory profiles, different mechanical constraints, and a different relationship to growth. The fetus is supposed to be building itself. The adult is supposed to be maintaining itself. Recreating fetal-like healing in an adult does not mean making the adult temporarily "more fetal" in a poetic sense; it means altering immune signaling, extracellular matrix composition, and mechanical forces in ways that might have systemic consequences.

The more ambitious regenerative technologies also raise questions about enhancement. If one can reduce scarring after injury, could one reduce scarring after elective surgery? If one can improve cartilage repair, could one make joints that outperform their natural baseline? If one can stimulate muscle regeneration, could one maintain strength far beyond what aging would otherwise permit? Enhancement is often framed as a separate domain from therapy, but regeneration blurs the line. A regenerated tissue might not match the original; it might be stronger, or at least tunable. Even the choice of what counts as "normal" becomes negotiable when the body is modifiable.

A practical moral problem follows: who gets to decide acceptable risk in exchange for potential restoration? For a person with a spinal cord injury, a small increase in long-term cancer risk might be worth accepting if it offers even modest functional recovery. For a person seeking scar reduction for cosmetic reasons, the same risk might be unacceptable. Yet the biological mechanism might be similar. Medicine will need not only technical targeting but ethical targeting: policies and norms that match the intensity of intervention to the seriousness of need.

None of this implies that regenerative therapies should be feared into paralysis. The history of medicine is full of interventions that were once considered dangerously radical. Organ transplantation itself required immunosuppression, increasing infection and cancer risks, and yet it became ethically and clinically acceptable because the benefits were enormous and the risks could be managed and communicated. The same is true of chemotherapy, radiation, and many biologic drugs that modulate immunity. The difference is that regeneration targets the basic permissions that keep cells civilized.

That civility is enforced by multiple layers of control: cell-intrinsic checkpoints that monitor DNA integrity, immune surveillance that removes aberrant cells, tissue architecture that limits expansion, and systemic signals that coordinate growth with the body's overall state. A regenerative therapy that works well will almost certainly borrow from those controls rather than attempting to bypass them. It will aim for local, temporary plasticity with rapid restoration of normal governance. The most convincing future therapies are likely to be those that include their own exit strategy: a built-in return to stability, not merely the hope that stability will reassert itself.

If the myth of Prometheus is invoked here, it is not because regeneration is a forbidden fire that ought to remain untouched. It is because the gift, once given, changes the relationship between humans and their limits. Fire cooks food and forges tools, but it also burns cities. A regenerative medicine capable of overriding scar would rebuild bodies and also rearrange how society thinks about injury, aging, work, and responsibility. The question is not whether the technology will be used. The question is whether it will be used in a way that respects the long timescales of biology and the uneven landscape of human life.

The most honest vision of regenerative medicine is neither utopian nor dystopian. It is a clinic where repair is possible but never free: paid for in careful targeting, in long monitoring, in strict boundaries around who should receive what, and in humility about how much plasticity a human body can safely tolerate. The body can be coaxed toward rebuilding, but it will always insist on asking, in its molecular language, the same cautious question: *and how do we make sure it stops?*

www.ingramcontent.com/pod-product-compliance
Lightning Source LLC
LaVergne TN
LVHW022210200926
848481LV00006B/1916